Seven Deadliest Wireless Technologies Attacks

Syngress Seven Deadliest Attacks Series

Seven Deadliest Microsoft Attacks
ISBN: 978-1-59749-551-6
Rob Kraus

Seven Deadliest Network Attacks
ISBN: 978-1-59749-549-3
Stacy Prowell

Seven Deadliest Social Network Attacks
ISBN: 978-1-59749-545-5
Carl Timm

Seven Deadliest Unified Communications Attacks
ISBN: 978-1-59749-547-9
Dan York

Seven Deadliest USB Attacks
ISBN: 978-1-59749-553-0
Brian Anderson

Seven Deadliest Web Application Attacks
ISBN: 978-1-59749-543-1
Mike Shema

Seven Deadliest Wireless Technologies Attacks
ISBN: 978-1-59749-541-7
Brad Haines

Visit **www.syngress.com** for more information on these titles and other resources

Seven Deadliest Wireless Technologies Attacks

Brad Haines

Technical Editor **Tim Kramer**

AMSTERDAM • BOSTON • HEIDELBERG • LONDON
NEW YORK • OXFORD • PARIS • SAN DIEGO
SAN FRANCISCO • SINGAPORE • SYDNEY • TOKYO
Syngress is an imprint of Elsevier

ELSEVIER

SYNGRESS®

Syngress is an imprint of Elsevier.
30 Corporate Drive, Suite 400, Burlington, MA 01803, USA

This book is printed on acid-free paper.

Library of Congress Cataloging-in-Publication Data
Application submitted

British Library Cataloguing-in-Publication Data
A catalogue record for this book is available from the British Library.

ISBN: 978-1-59749-541-7

Printed in the United States of America
10 11 12 13   5 4 3 2 1

For information on rights, translations, and bulk sales, contact Matt Pedersen, Commercial Sales Director and Rights; e-mail: m.pedersen@elsevier.com

For information on all Syngress publications, visit our Web site at www.syngress.com

Typeset by: diacriTech, Chennai, India

Contents

A preview chapter from *Seven Deadliest Network Attacks* can be found after the index.

Acknowledgments

I would like to acknowledge all the people who have helped me over the years to gain my knowledge of wireless and provided me with a fun and interesting hobby and career and the opportunity to share that knowledge with others.

- Mike Kershaw (Dragorn) for putting up with my constant feature requests and bug reports for Kismet and for making his wonderful tool free to the world.
- Frank Thornton (Thorn) for providing mentorship and camaraderie at many a convention and for helping design some great wireless contests to hone my skills.
- Chris Hurley (Roamer) for many of the wireless contests and for years of direct and honest opinions and comments on everything wireless.
- Josh Wright for his wonderful work on coWPAtty and the WPA cracking tables and for many other tools used by thousands of people on a daily basis.
- Emmanual Goldstein, 2600 magazine, and the HOPE organizers for another wonderful conference full of new ideas and experiences and giving me the opportunity to meet my wife.
- Jeff Moss (Dark Tangent) and the DEFCON staff for continually putting on a wonderful conference and providing a great environment to learn and explore new things and ideas.
- The DEFCON and Netstumbler forum regulars who have provided years of insight, tutelage, and, at times, questionable encouragement in my research.
- The "Church of WiFi" members for helping on so many projects and for keeping wireless sexy.
- Jesse Burns, Simple Nomad, Adam Laurie, Zac Franken, and all the others who have helped over the years.
- Finally, and most importantly, my wife Dianna (Grey Frequency) for always being there for me no matter how weird, outlandish, or crazy the project and for being a voice of reason when I need it.

About the Authors

Brad "RenderMan" Haines Contributing Author to *RFID Security* (ISBN: 978-1-59749-047-4, Syngress) and *Kismet Hacking* (ISBN: 978-1-59749-117-4, Syngress), is chief researcher of Renderlab.net and co-refounder of "The Church of WiFi" wireless thinktank. He currently operates his own consulting company in Edmonton, Alberta, Canada, providing wireless performance and security assessment services to a variety of clients both large and small.

A noted expert in the hacker community in the field of wireless security, he has spoken at many international conferences such as Black Hat and DEFCON and taught several classes on free wireless assessment tools. He has also contributed over time to many wireless security tools such as the Kismet wireless sniffer and coWPAtty.

Technical Editor

Tim Kramer (CISSP, CEH, GSEC, GCIH, NSA IAM/IEM) recently served as subject matter expert (SME) for Information Assurance (IA), Unix systems, and wireless technologies for Honeywell Technology Solutions and General Dynamics. Currently, he is providing IA and Unix support for various organizations in Portsmouth, Virginia. Tim's background includes positions such as vulnerability analyst at NETWARCOM (Naval Network Warfare Command) and roles such as the Navy's wireless SME, coauthor of Navy and Joint Forces wireless policies, and IA review of one of Electronic Data System's wireless solutions.

Introduction

INFORMATION IN THIS CHAPTER

- Book Overview and Key Learning Points
- Book Audience
- How This Book Is Organized

Look around any street, business, or home. There is a world of information passing in front of us and through us that many people just don't perceive. Wireless signals are everywhere, passing information between sources at incredible speeds just beyond our vision. We all use dozens of wireless devices in our daily lives, simple things like wireless doorbells that communicate very simple information to modern smart phones that have access to all the knowledge of the Internet. It's when you start to realize the amount of information that is flowing freely past our eyes and that with the right tools, you can see that information that the need for security becomes apparent and the value it has to an attacker.

If you can see this information, who else may be able to?

This book attempts to show the implications of this brave new wireless world with seven major attacks against them along with how best to protect yourself and your private information. From Wi-Fi networks and cordless phones to RFID and good old analog, there is no end to the threats from wireless devices.

BOOK OVERVIEW AND KEY LEARNING POINTS

Each chapter in this book covers an example of a failure of wireless technology. Sometimes, it is a specific and familiar technology, and others are more about specific concepts that cover multiple technologies. It is important to learn the implications of deploying wireless devices in your home or business, and each chapter starts with an example of a danger associated with wireless. From there, workings of the attack works are discussed so that you can see things from an attacker's view, to help you better understand the risks and how best to mount a defense. From there, options for defense are discussed with associated strengths and possible weaknesses. Wireless security is tricky and, short of not using wireless, there is no one single solution to this very complex problem.

Information is a commodity; it can be bought and sold just like any tangible product. Attackers are already well aware of the ease and availability at which wireless signals can yield valuable information for identity theft, fraud, industrial espionage, blackmail, and to facilitate other types of crime, both high and low tech.

BOOK AUDIENCE

This book was written with the novice in mind, the small business IT guy who is trying to understand how best to secure things when his boss sends down a request for some new wireless piece of technology. This book is applicable to anyone in a technical role either as the family tech support to corporate IT managers. Since wireless affects us all, we all have know how to take steps to protect ourselves. While this book only covers seven specific issues, you will hopefully learn the tricks and tools needed to assess the security of the potentially millions of different wireless devices on the market and how they impact your security.

This book is assuming some familiarity with basic networking and computer use. Knowledge of radio is not required, but certainly worth exploring on your own if you find this fascinating.

HOW THIS BOOK IS ORGANIZED

This book contains seven chapters that address different attacks against popular wireless protocols and systems. Each chapter includes an example real attack scenario, an analysis of the attack, and methods for mitigating the attack. Common themes will emerge throughout the book, but each wireless technology has its own unique quirks that make it useful to attackers in different ways, making understanding all of them important to overall security as rarely is just one wireless technology in use at a home or office.

Chapter 1: 802.11 Wireless – Infrastructure Attacks

The ubiquitous 802.11 wireless network is covered first. It's hard to go anywhere without running across this type of network. It has become an invaluable resource for both home and office for networking and Internet access. Wireless networking is also incredibly valuable to attackers as it gives the attacker opportunity to access networks at a safe distance, almost as if they were connected to the wired network. Chapter 1 focuses on the infrastructure of these networks and the security implications of their use and how to and how to not secure them. They may be ubiquitous, but that doesn't mean they are secure.

Chapter 2: 802.11 Wireless – Client Attacks

Wireless clients, those devices that talk to the rest of the wireless network, are covered in Chapter 2. Attackers, stymied by increasing amounts of security on the infrastructure side, are changing tactics and attacking client devices directly. At home or away, wireless clients and the information they contain and communicate are tempting targets for pranksters and thieves alike.

Chapter 3: Bluetooth Attacks

Bluetooth is the subject of Chapter 3. This common protocol was meant to replace cable clutter but has become so much more. While it is meant for short range, any distance can be a comfort for an attacker. Modern devices carry a great deal of information, tempting for a new era of digital pick pockets. You could lose everything without losing anything.

Chapter 4: RFID Attacks

RFID is a technology most people are not even aware of despite the billions of tags in use everyday. As the subject of Chapter 4, RFID is looked at with an eye to how its perceived benefits can actually be their greatest vulnerability and how they can be thwarted by those with ill intentions. RFID is all around us, and knowing how to identify it and how to protect it is a very important topic not often understood by many people.

Chapter 5: Analog Wireless Devices

Even the most modern of wireless devices often at their heart are still just radios. Often these new devices are using age-old radio techniques to allow their communication. Chapter 5 will show you how to identify these devices and understand the risks associated with their use and how vulnerabilities apparent over 100 years ago are still around to make life interesting.

Chapter 6: Bad Encryption

A common solution to wireless security problems is to add encryption. The common problem though with wireless security is bad encryption. Poor design choices, hardware limitations, and cost can all turn a good security idea into a failure at record speed. Chapter 6 looks at this problem with a number of real-world examples and shows how something that was supposed to protect communications can end up providing less security than advertised.

Chapter 7: Cell Phones, PDAs, and other hybrid devices

It's impossible to escape them, but cell phones are everywhere. Today's modern smart phones and other hand-held gizmo's are at their heart, computers in their own right and have their own unique security issues that need to be considered. Chapter 7 will look at these new generation devices and how their small size, portability, and communication capacity make them interesting and tempting targets for today and the future.

CONCLUSION

Wireless devices are here to stay, and anyone involved in IT or just life in general will end up interacting with wireless devices almost everyday. They are highly unlikely to go away and are more than likely to become more pervasive and more complex as time progresses. A firm understanding of the risks involved with deploying and using wireless devices in your life will help you protect against attacks and allow for all the benefits of the wireless world while minimizing the risk.

802.11 Wireless – Infrastructure Attacks

1

INFORMATION IN THIS CHAPTER

- How Wireless Networks Work
- Case Study: TJX Corporation
- Understanding WEP Cracking
- How to Crack WEP
- It Gets Better and Worse
- WPA and WPA2 in a Nutshell
- How to Crack WPA PSK and WPA2 PSK

Wireless is a term thrown about quite a bit lately. Everything seems to be wireless to one degree or another, even some things no one ever expected to be, like refrigerators and other appliances. Most often, when the term wireless is used in regards to computing, it's to do with 802.11 networks.

Just about every new laptop that hits the market today has an 802.11 network card built in. It's a technology that has become ubiquitous in our lives, and we can hardly remember a time when it wasn't part of our days. It's a technology that has grown in terms of speed and range to provide the capability to be connected to the Internet from anywhere in our homes or businesses.

This widespread technology would also very quickly become quite an issue from a security perspective. Users quickly demanded to "cut the cable" and be able to access the network from anywhere in the office. Home users were quick to adopt the technology to work from the kitchen, the couch, or (more oddly) the bathroom. This intense push led to a lot of overworked and underpaid information technology (IT) administrators and neighborhood computer know-it-alls to install wireless networks without properly understanding the security risks involved. These early networks would continue to "just work" with users not realizing that the security arms race caught up with them and even passed them, making them prime targets for attack.

In November 2003, Toronto, Ontario, police held a press conference to announce a (at the time) new and unusual crime.[A] The police report indicates that at around 5:00 A.M. an officer noticed a car slowly driving the wrong way down a one-way street in a residential neighborhood. The officer pulled the car over, and when he walked up to the driver, he was greeted with several disturbing sights. The driver was first of all not wearing any pants, which is probably disturbing in and of itself, but more alarmingly, on the passenger seat was a laptop clearly displaying child pornography. The driver had been using open wireless networks in the area to obtain Internet access to download child pornography, unbeknownst to the owners of those networks. The owners were victims themselves, twice. First, they were victims of theft of service since their communications had to compete for bandwidth with the traffic of the unauthorized user. Second, they were victimized because, for all intents and purposes, the child pornography was being downloaded through their connection. Any digital trail left would lead back to them, potentially exposing them to false accusations of downloading child pornography themselves and all the emotional and financial damage that accusation can bring. The suspect's home was searched as a result, and 10 computers and over 1,000 CDs worth of illegal material were seized.[B]

This case, along with others through the years, has shown that operating an access point (AP) without any authentication of client devices is dangerous. If anyone can connect, there is no restriction on what sort of activities those users can partake in. Often, it's simply to check an e-mail or catch up on the latest news, but it may be someone downloading copyrighted materials, sending threatening messages, or doing worse.

Sometimes, connecting to an open network without authorization can occur even without someone realizing he or she is doing it. Windows XP, before Service pack 2, was notorious for automatically connecting to networks named the same as ones it had connected to before. A person carrying a laptop down the street configured for a common network name like "linksys" could drift to any network similarly named "linksys" and be committing an unauthorized access without knowing or interacting. Many users noticed this behavior and thought it more than helpful in gaining access to free Wi-Fi. Attackers noticed this and began to exploit it (more on that in Chapter 2, 802.11 Wireless – Client Attacks).

It's sad to consider that leaving your APs open for anyone to connect to is a dangerous proposition. The idea of everyone sharing free Internet access anywhere he or she goes is a tempting one, but society, as a cross section, contains all sorts of people, some good and some bad, and often the bad ruin such freedoms for everyone.

The Institute of Electrical and Electronics Engineers (IEEE) knew that they had to establish some mechanism to maintain privacy of communications as they were broadcast and restrict who can connect and from where. This is why all APs sold contain various methods of securing communications and limiting who can connect.

[A] www.ctv.ca/servlet/ArticleNews/story/CTVNews/1069439746264_64848946/?hub=CTVNewsAt
[B] See http://citeseerx.ist.psu.edu/viewdoc/summary?doi=10.1.1.116.850 and click on the PDF icon underneath the cached link on the upper-right side of the page.

Originally, Wired Equivalent Privacy (WEP) was the only option available, but as time went on, Wi-Fi Protected Access (WPA) was introduced as an interim solution when WEP was shown to be weak, and eventually WPA2 was brought forth with the final ratification of 802.11i.

As with many security technologies, if you give users the option of using it, they often won't. If you give them too many options, there's no way of guaranteeing that they will keep their systems up to date either.

HOW WIRELESS NETWORKS WORK

A wireless network typically is made up of two classes of device: APs and client devices, typically called *stations* (STAs). This chapter focuses on security of APs typically found in a home or business. Client security is discussed in Chapter 2, 802.11 Wireless – Client Attacks. These networks can be 802.11a, b, g, or n, but for the most part, and for discussion purposes in this chapter, it doesn't matter. The infrastructure needed is fairly universal, and standards for security are pretty much the same for all of them.

The APs are something everyone in the IT industry and most home computer users are probably familiar with. They come in all shapes and sizes and can have varying features. They are the gateways between the wired and wireless network. If you don't have one at home already, you can usually see them bolted to the wall at many businesses or in public spaces with one or more antennas sticking out of them. The AP is what the client STA connects to in a wireless network (as opposed to the other way around). In their default state, most APs will accept connections from any client STA that asks to join the network. While this is convenient for users, it is also very convenient for anyone else who wants to connect, for good reasons or bad.

In the early days of wireless, this was seen as something positive. Wireless brought out ideas of a brave new world with free Internet access and sharing of a new and useful resource. It didn't take long for the bad guys to figure out that this was very useful for them as well.

NOTE

It's hard to imagine a world without wireless networking. It's absolutely everywhere. Since 2001, Wigle.net, an online repository of data submitted by users, has collected tens of millions of unique network locations with Global Positioning System (GPS) coordinates and over a billion points of observations of those networks. The site also includes some automatically generated maps of that data that can pretty conclusively show that wherever there are people and computers, there are wireless networks. Figure 1.1 shows Wigle.net's map of North America.

While this sort of activity may seem odd, companies like Skyhook Wireless (www.skyhookwireless.com) has made a business out of wardriving themselves. They map the location of networks throughout the world and use that information to provide GPS-like location sensing via triangulation of known APs as opposed to satellites, which has the added benefit of working indoors in many cases, unlike GPS.

As you can see, there are wireless networks everywhere. Wherever there is a population center, you will be able to find wireless networks there.

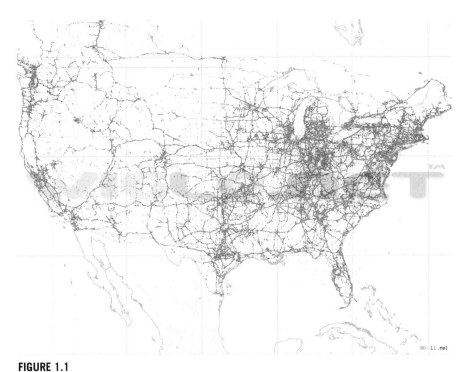

FIGURE 1.1

Wigle.net's Map of North America

Wireless is a shared medium. If you remember the bad old days where Ethernet networks were all using hubs and not switches, everyone saw everyone else's traffic. Well, wireless brings all the fun of those networks back. In those days, hubs were simple rebroadcasters, and they had no real intelligence as to what was connected to each port. A client would put a packet onto the wire and the hub would rebroadcast that packet to every other computer on the hub. The intended recipient accepts the packet, whereas the other simply ignores it. As you can imagine, have many clients on the network trying to communicate simultaneously and it gets pretty noisy. Network adapters normally filter out packets that come down the wire that are not intended for their address. If you disable that filter, you can now listen to all the packets, even the ones not intended for that network card. This is usually called *promiscuous mode* and has been a fundamental tool of network diagnosis since the beginning of networks.

In a wireless network, promiscuous mode does the same thing if you are associated to a network. If you want to listen to other networks without associating or get the management traffic at Layers 1 and 2, then you need to remove the filters from Layers 1 and 2 and the logical separation of networks. This is where monitor mode comes in. Monitor mode is useful as it allows the card to listen to any wireless data, from any network on the same channel in range.

An 802.11 network typically sends out advertising "beacon" frames to announce its presence. These frames contain the network BSSID (Media Access Control [MAC] address), ESSID (commonly known as SSID, the logical name for the network), and various flags about its capabilities (speed, encryption level, and so on). All this information is sent in clear text. Since it's a shared medium, anyone can pick up these beacons and this basic information. This is the essence of wardriving.

Much has been written about wardriving, but the best definition was coined on the netstumbler.org forums by a poster named blackwave:

> *Wardriving (v.) – The benign act of locating and logging wireless access points while in motion.*[1]

Essentially, it is using a wireless-enabled device to search for others. This can be as simple as the Wireless Zero Config utility in Windows searching for a network to programs like Kismet, a full wireless detector and sniffer. Integrating a GPS into the system, and coordinates of those networks can be used to generate maps of local areas for reports, or submitted to sites like wigle.net to add to a larger community data pool.

CASE STUDY: TJX CORPORATION

In April 2007, U.S. retail giant TJX, owners of TJ Maxx, Marshalls, and other retail store brands, publicly admitted in their annual Security and Exchange Commission filings that their network security had been breached and that customer credit card numbers and other information had been available to criminals roaming the network for over a year. The fallout for the company is expected to top 1 billion dollars over 5 years and caused headaches for millions of consumers now open to identity theft and credit card fraud, as well as credit card companies and financial institutions having to pay millions to replace consumers, credit cards. In May 2008, authorities arrested Albert Gonzalez in Miami, Florida, related to another large-scale identity theft. He was eventually charged as the ringleader in the TJX attacks and several other large corporate penetrations, and on August 28, 2009, Gonzalez agreed to a plea bargain and stands to serve 15 to 25 years for his role. There are several other outstanding charges related to similar attacks on other corporations that, at the time of this writing, are still waiting to work their way through the courts.

EPIC FAIL

An interesting note about Gonzalez is that it appears he began the attacks on TJX while working as an informant for the U.S. Secret Service.

Gonzalez was a member of the Shadowcrew, an online group that ran a Web site with over 4,000 members, devoted to the buying, selling, and trading of stolen credit card numbers. The group trafficked more than 1.5 million credit card numbers. When members of the group began to be arrested, Gonzalez turned informant and helped with the indictment of 19 other members of the Shadowcrew.

As he worked for authorities, he apparently began a new crime spree under the noses of his FBI handlers. This included the TJX attacks. An obvious black eye for the agency, his ability to hide his activities for so long is a useful lesson to future investigators. His usage of remote computers and encryption should be viewed as a testament to the creativity of online criminals to hide their activities.

While many details are not fully known, the seemingly biggest and most well-reported entry point was a St. Paul, Minnesota store's wireless network. The indictment[C] of Gonzalez and others indicates that Marshalls and TJX stores were penetrated through wireless networks in Miami from their own parking lots. The full extent may never be known, but it is clear that wireless networks were a component in these attacks.

Using freely available software, the attackers identified the network and proceeded to crack the WEP key used to secure the network. This provided access to the store's network and gave a foothold into the larger corporate network and all the data it contained. Whether it was a targeted attack of this specific store and chain, or if it was just that they happened by and noticed the weak security, we probably won't know. Various prosecutions of the perpetrators, though, show that many different companies were penetrated and were probably all just targets of opportunity rather than of a specific agenda. The one common element seems to be the presence of these businesses along U.S. interstate 1 in Florida. Likewise, the attackers just drove down the interstate and collected data, returning to tempting and weak targets later.

Various reports since then have indicated that the store's wireless network was secured using WEP. At the time, WEP was known to be fatally flawed and was already outmoded by the introduction of WPA encryption. These networks are often installed for the convenience of bar-code-reading scanner guns used at many stores for inventory control; these connect back to the store server over wireless. Many of these systems are only capable of WEP and are non-upgradeable, and given the amount already invested, companies are often slow to upgrade. Further complicating matters and contributing to the complacency was that, at the time, stores had to meet the Payment Card Industry (PCI) security standards in order to be allowed to take credit and debit cards. Recommendations were made to TJX to upgrade its wireless security to WPA; however, it seems from corporate e-mails that upgrades were delayed in favor of the cost savings associated with not replacing the equipment in

---

[C]www.justice.gov/usao/ma/Press%20Office%20-%20Press%20Release%20Files/IDTheft/
Gonzalez,%20Albert%20-%20Indictment%20080508.pdf

many stores. In addition, VISA, one of the members of the PCI group, gave TJX a pass on their compliance with the condition they would do something to improve their wireless security in time. One can be certain that after the incident, wireless security was taken much more seriously. Suddenly, the original costs of upgrading seem a lot smaller than the subsequent costs of cleanup and bad press.

UNDERSTANDING WEP CRACKING

WEP was the original encryption scheme included in the 802.11b wireless standard from 1997. At the time, strong encryption was considered a defense by the U.S. State Department (a lot of manufacturers' head offices were located in the United States) and since there were restrictions on exportation of strong encryption to foreign countries, the key length was limited to 40 bits. This was later relaxed to allow 64- and 128-bit keys to be exported. For many years, this was the only security standard available for wireless.

> **NOTE**
>
> There have been many proprietary security methods available as well. Some, such as Lightweight Extensible Authentication Protocol (LEAP), are better than WEP but require end-to-end solutions from a single company like Cisco. This increased cost and broke much of the interoperability that made 802.11 so appealing, and never caught on outside homogenous networks in corporations and almost never for home users.

As early as 2001, implementation problems with the WEP encryption scheme led to the first real break. The problem revolved around the initialization vector (IV) field of the scheme, a random number concatenated with the network key, used to provide some randomization to the scheme. WEP is based on the RC4 stream cipher algorithm, and as with any stream cipher, identical keys must not be used. The IVs change with each packet and eventually repeat, giving an attacker two packets with identical IVs. The counter used for IVs was 24 bits long, which on a fairly busy network meant that there was a good chance that after 5,000 packets, an IV would be repeated, yielding an IV collision where two packets were encrypted with the same key, thus providing a basis for cryptanalysis. If more collisions are encountered, this increases the chances of an attack.

Tools began to emerge like Airsnort[D] that required 5 to 10 million packets to be captured for analysis. On a particularly busy network, this would take a couple of hours to collect. On quieter networks, it could take days, and even then, it was very much a hit-or-miss situation. These tools were later replaced by the original Aircrack[E] suite of tools, which introduced some new methods of attack and reduced the amount of data needed between 200,000 and 500,000 packets for 40- and 64-bit WEP and a million for 128-bit WEP, a much more manageable amount to capture.

[D]http://airsnort.shmoo.com/
[E]www.aircrack-ng.org/

Further development of tools allowed for faster and more efficient use of IV data. The advent of the ARP replay attack really shortened the time needed to perform an attack. The ARP replay attack is where an encrypted ARP packet (known because of its unique size, even when encrypted) is captured from a network and retransmitted back to the AP, which in turn sends back another ARP packet with a different IV. This is done rapidly and repeatedly and creates a huge amount of IVs to be used and the counter to roll over and duplicate IVs to be sent. This, along with improvements to Aircrack (by this time abandoned by the original author and now reimplemented as Aircrack-ng), reduced the time to execute an attack from hours and days to as little as 10 min.

The Pychkine–Tews–Weinmann (PTW) attack was arguably the final nail in the coffin for WEP. This attack was able to use more of the packets for analysis and only needed 20,000 to 50,000 packets to work. In combination with the ARP replay attack, this could be executed in as little as 60 s, start to finish, yielding the hexadecimal WEP key for the target network.

In the case of TJX at the time of the initial attack, it was widely known WEP had issues and some of these tools had been around for a few years already (since at least 2005 for Aircrack). It was just a matter of a determined attacker to spend the necessary time and energy along with a laptop, wireless card, and felonious intention to penetrate the wireless network at that fateful store one night.

HOW TO CRACK WEP

Cracking WEP today is actually a frighteningly easy prospect. Original tools were fairly slow, hit or miss, and generally required a lot of data. After approximately 7 years of development, these tools have reached a point of refinement that makes breaking WEP a fairly reliable outcome.

There are many tools available that break WEP, but the most popular is Aircrack-ng ("ng" denoting new or next generation as opposed to the original Aircrack by Chris Devine that has been since abandoned). This section will be a quick tutorial to the steps necessary to break WEP and recover a key.

WARNING

Under most jurisdictions, any attempt to recover a key from a network you do not own or have permission to do so is very likely a crime. As noted with the TJX attack, there are very real consequences to breaking networks, no matter how easy it may seem.

If you want to test this or any other security tool, it is the best and safest thing to do so with your own equipment or that which you have express permission from the owner. Cracking your neighbor's WEP key may sound like fun, but it's a felony. Don't do it.

This guide assumes several things since there is no way to know what exact configuration of equipment you might be using. Usage should be similar no matter your platform:

- Laptop running Ubuntu 9.04
- Atheros 802.11b/g wireless card
- Madwifi-ng drivers (0.9.4) from http://madwifi-project.org/

> **TIP**
>
> Probably the most important part of any wireless tool kit will be a compatible wireless card. For the most part, Atheros-based work is the best for Linux and wireless penetration testing since the drivers for Linux are very open and capable of doing many of the odd things necessary to enable some of these attacks.
>
> Check the Aircrack-ng Web site for a list of supported and recommended cards: www.aircrack-ng.org/doku.php?id=compatibility_drivers

The first step is to acquire the Aircrack software. This is available from the Web site www.aircrack-ng.com. It is available in several different packages for Unix systems and Windows along with bootable Linux distros and VMware images.

Aircrack is actually a suite of tools. The namesake is the actual tool that does the cracking. Around it are many helper applications to help you get what you need. For the most part, you will only need to worry about four programs: Airmon-ng, Airodump-ng, Aireplay-ng, and Aircrack-ng.

Once you have the Aircrack suite installed or built, you'll need to start capturing packets. To do so, you'll need to put the card into monitor mode in order to listen to packets.

> **TIP**
>
> For the Madwifi drivers used in making this guide, it was necessary to take some additional steps to prepare the card for monitor mode.
>
> The Madwifi drivers operate as a "parent" device usually named wifi0, wifi1, etc, and make virtual interfaces (VAPs) that are what programs actually interact with (ath0, ath1, etc). You can have up to four VAPs off of any one parent device acting in different capacities but having multiple devices can cause issues. For instance, if you have one VAP in monitor mode trying to channel hop and another in client (STA) mode, if the STA-VAP associates with an AP, the parent device cannot keep changing channels, which screws up the monitor mode VAP.
>
> Unless you need to have multiple VAPs, it's a good idea to destroy them before creating a new one to do monitor mode. This is easily done with the command *wlanconfig*:
>
> ```
> wlanconfig ath0 destroy
> ```
>
> (where ath0 is the VAP you want to destroy). Otherwise, if you want to start fresh, you can unload and reload the Atheros driver and force it not to create any VAPs:
>
> ```
> modprobe -r ath_pci
> modprobe ath_pci autocreate=none
> ```

Assuming in this case that you have a compatible card ready to go into monitor mode, the Airmon-ng program makes it easy to set your card. The program identifies the driver and knows what steps to take to enable monitor mode. In the case of Madwifi drivers, specify the parent device, wifi0, as the interface:

```
airmon-ng start wifi0
```

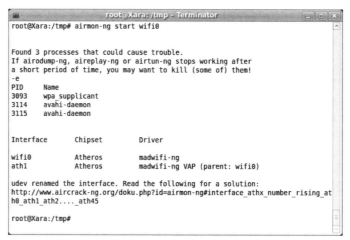

```
root@Xara:/tmp – Terminator
root@Xara:/tmp# airmon-ng start wifi0


Found 3 processes that could cause trouble.
If airodump-ng, aireplay-ng or airtun-ng stops working after
a short period of time, you may want to kill (some of) them!
-e
PID     Name
3093    wpa_supplicant
3114    avahi-daemon
3115    avahi-daemon


Interface      Chipset       Driver

wifi0          Atheros       madwifi-ng
ath1           Atheros       madwifi-ng VAP (parent: wifi0)

udev renamed the interface. Read the following for a solution:
http://www.aircrack-ng.org/doku.php?id=airmon-ng#interface_athx_number_rising_at
h0_ath1_ath2.... ath45

root@Xara:/tmp#
```

FIGURE 1.2

Airmon-ng's Output for a Madwifi-ng Card

Airmon-ng will report the name of the interface in monitor mode and look something like Figure 1.2.

Once the card is in monitor mode, you can collect packets. The program in the suite to do this is called *airodump-ng*. Many options can be set from the command line and are specified before the interface name at the end:

```
airodump-ng --channel 3 --write foo ath1
```

This command specifies Airodump to listen on Channel 3 (as opposed to hopping through all channels), write out the captured data to a file with the prefix "foo," and use interface ath1, which is our monitor mode interface from Airmon-ng. If you are interested in the other options for Airodump or any program in the suite, just run it with the -help switch.

Once it is running, you should see networks begin to populate the columns and packet counts start to rise. Figure 1.3 shows a single network being captured; however, if you are in a noisy area, more networks may show up.

If everything is running fine, you should see one or more networks listed in the display. Hopefully, your target is one of them. The upper portion lists currently active APs and some basic information about them. The lower portion lists client devices and their associations.

Depending on the amount of traffic already on the network, you may start seeing the data column number for your target that start to rise. If it is rising fast enough to acquire 50,000 packets in what you deem a reasonable amount of time, then it's just a matter of time to wait. If there is no traffic, or no clients, or you are just impatient, you can try an ARP injection attack to force more data to be generated.

Aireplay-ng handles most of the active attacks for WEP. It should be run at the same time as your capture program, such as Airodump-ng. It has many command-line

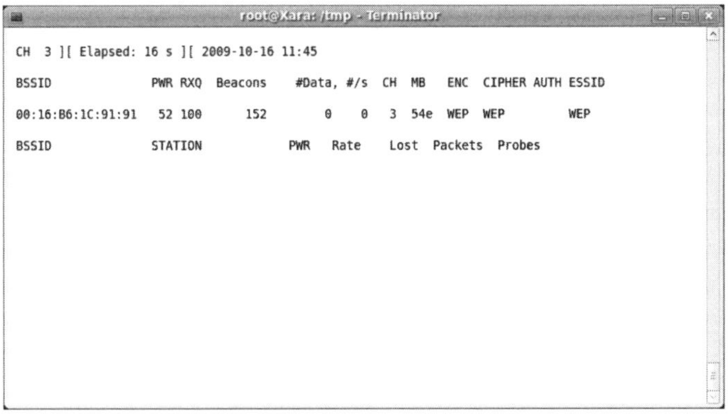

FIGURE 1.3

Airodump-ng's Output Capturing Data from Our Target Network Named WEP

switches, but for the purposes of this guide, we will focus on the ARP replay as it is the most effective and widespread:

```
aireplay-ng -3 -b 00:16:B6:1C:91:91 ath1
```

The above command is specifying Aireplay-ng should use attack number 3, which is the ARP replay attack, the BSSID of the target, and the interface to listen and inject on. Since we didn't specify a "-h" or host option, it defaults to using the local interface MAC address. For better results, if you can, specify the MAC for an already associated client. Be careful as some cards are incapable of injection in monitor mode and may need special drivers or preparation.

NOTE

For whatever reason, the Aircrack-ng developers decided that for some attacks "-a" would be the switch to specify the BSSID of the target, for others it's "-b." If you use the wrong one, Aireplay-ng is smart enough to alert you as to which one to use instead.

Aireplay-ng will listen on the interface for an ARP packet. ARP packets have a specific size (68 or 86 bytes) even when encrypted and stand out. Once it captures one, it will retransmit it back to the AP causing a reply with another ARP encrypted with a different IV and Aireplay-ng will begin counting up ARP packets as demonstrated in Figure 1.4. This will start a flood of data to come in to run the crack on.

Should you be attacking an AP without any clients and using your local MAC for injection doesn't seem to be working, you can sometimes force an ARP packet to be generated by faking an association to the AP from your attacking client.

```
root@Xara: /tmp - Terminator                                      _ □ x
root@Xara:/tmp# aireplay-ng -3 -b 00:16:B6:1C:91:91 ath1
No source MAC (-h) specified. Using the device MAC (00:16:CF:67:C0:47)
11:56:06  Waiting for beacon frame (BSSID: 00:16:B6:1C:91:91) on channel 3
Saving ARP requests in replay_arp-1016-115606.cap
You should also start airodump-ng to capture replies.
Read 12862 packets (got 140 ARP requests and 0 ACKs), sent 135 packets...(499 pp
Read 13049 packets (got 201 ARP requests and 0 ACKs), sent 185 packets...(499 pp
Read 13238 packets (got 259 ARP requests and 0 ACKs), sent 235 packets...(499 pp
Read 13407 packets (got 322 ARP requests and 0 ACKs), sent 286 packets...(501 pp
Read 13618 packets (got 382 ARP requests and 0 ACKs), sent 336 packets...(500 pp
Read 13799 packets (got 441 ARP requests and 0 ACKs), sent 385 packets...(499 pp
Read 13987 packets (got 500 ARP requests and 0 ACKs), sent 435 packets...(499 pp
Read 14179 packets (got 560 ARP requests and 0 ACKs), sent 485 packets...(499 pp
Read 14368 packets (got 623 ARP requests and 0 ACKs), sent 536 packets...(500 pp
Read 14556 packets (got 681 ARP requests and 0 ACKs), sent 586 packets...(500 pp
Read 14760 packets (got 741 ARP requests and 0 ACKs), sent 636 packets...(500 pp
Read 14971 packets (got 807 ARP requests and 0 ACKs), sent 685 packets...(499 pp
Read 15149 packets (got 867 ARP requests and 0 ACKs), sent 736 packets...(500 pp
Read 15350 packets (got 929 ARP requests and 0 ACKs), sent 786 packets...(500 pp
Read 15556 packets (got 991 ARP requests and 0 ACKs), sent 836 packets...(500 pp
Read 15765 packets (got 1057 ARP requests and 0 ACKs), sent 886 packets...(500 p
Read 15965 packets (got 1117 ARP requests and 0 ACKs), sent 936 packets...(499 p
Read 16166 packets (got 1185 ARP requests and 0 ACKs), sent 986 packets...(499 p
Read 16383 packets (got 1248 ARP requests and 0 ACKs), sent 1036 packets...(499
```

FIGURE 1.4

Aireplay-ng's Output for a Successful ARP Injection

```
root@Xara: /tmp - Terminator                                      _ □ x
root@Xara:/tmp# aireplay-ng -1 3 -a 00:16:B6:1C:91:91 ath1
No source MAC (-h) specified. Using the device MAC (00:16:CF:67:C0:47)
11:58:39  Waiting for beacon frame (BSSID: 00:16:B6:1C:91:91) on channel 3

11:58:40  Sending Authentication Request (Open System)
11:58:40  Authentication successful
11:58:40  Sending Association Request
11:58:40  Association successful :-) (AID: 1)
11:58:43  Sending Authentication Request (Open System)
11:58:43  Authentication successful
11:58:43  Sending Association Request
11:58:43  Association successful :-) (AID: 1)
11:58:46  Sending Authentication Request (Open System)
11:58:46  Authentication successful
11:58:46  Sending Association Request
11:58:46  Association successful :-) (AID: 1)
11:58:49  Sending Authentication Request (Open System)
11:58:49  Authentication successful
11:58:49  Sending Association Request
11:58:49  Association successful :-) (AID: 1)
11:58:52  Sending Authentication Request (Open System)
11:58:52  Authentication successful
11:58:52  Sending Association Request
11:58:52  Association successful :-) (AID: 1)
```

FIGURE 1.5

Aireplay with a Successful Association

While the ARP attack continues to run, run Aireplay in another terminal but this time using attack number# 1, the fake association:

```
aireplay-ng -1 3 -a 00:16:B6:1C:91:91 ath1
```

This command specifies attack number 1 (association) with a delay of 3 s between attempts against the specified BSSID using interface ath1 to listen and inject on. Since no "-h" option was set, the local MAC is used. These will then try and associate to the AP and hopefully an ARP packet will be generated to bootstrap ARP injection. A successful association looks something like Figure 1.5, including the smiley faces.

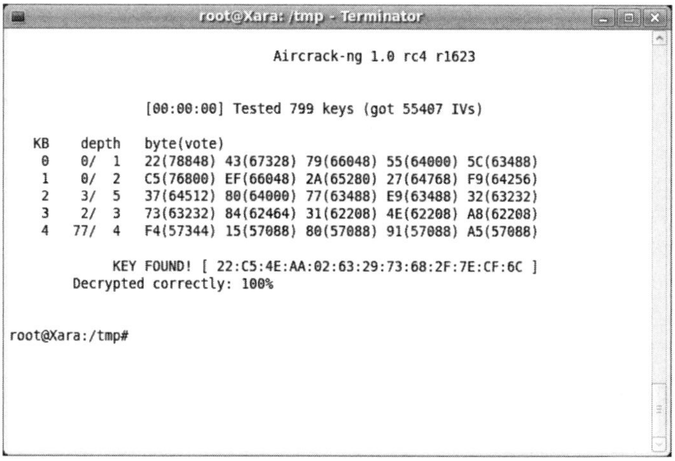

```
                    root@Xara: /tmp - Terminator

                   Aircrack-ng 1.0 rc4 r1623


             [00:00:00] Tested 799 keys (got 55407 IVs)

   KB    depth   byte(vote)
    0    0/  1   22(78848) 43(67328) 79(66048) 55(64000) 5C(63488)
    1    0/  2   C5(76800) EF(66048) 2A(65280) 27(64768) F9(64256)
    2    3/  5   37(64512) 80(64000) 77(63488) E9(63488) 32(63232)
    3    2/  3   73(63232) 84(62464) 31(62208) 4E(62208) A8(62208)
    4   77/  4   F4(57344) 15(57088) 80(57088) 91(57088) A5(57088)

             KEY FOUND! [ 22:C5:4E:AA:02:63:29:73:68:2F:7E:CF:6C ]
          Decrypted correctly: 100%


 root@Xara:/tmp#
```

FIGURE 1.6

Aircrack-ng Found the Key

Once Aireplay-ng is injecting, the Airodump data column total for the target should start to rise at a great speed. An optimal setup will only take a minute or two to approach the necessary starting point of 10 to 20,000 packets. You can collect more, but why not start early?

Airodump continually writes to the specified storage file, which in the guide is "foo-01.cap."

Aircrack-ng is run against this file and the packets for various networks are parsed:

```
aircrack-ng foo-01.cap
```

If there are packets for more than one network, you will be prompted to specify which one by selecting it from a list. Once that is done, the attack runs. First it tries the PTW attack, then the earlier analysis of the data in an effort to retrieve the key.

A successful attack should yield results like Figure 1.6 but with a different key.

If no key is retrieved or there are not enough packets, you can quit the program and run it later, or just leave it while more are captured and it will retry at periodic intervals.

Once Aircrack-ng gets a key, you can shut down the other programs and use the key to connect or do whatever you want with it.

IT GETS BETTER AND WORSE

Over time, people have slowly started to get the message that they need to secure their networks. Security experts and the wardriving community took every opportunity to warn people of the dangers of operating open networks.

> **NOTE**
>
> Interestingly, wigle.net, in their graph of crypto usage over time from submitted networks, shows a marked "jump" of almost 1 percent centered on the period of April 2007 when TJX announced they had a problem and the subsequent media attention to the issue of wireless security. Figure 1.7 shows the graph of encryption usage from wigle.net over time.

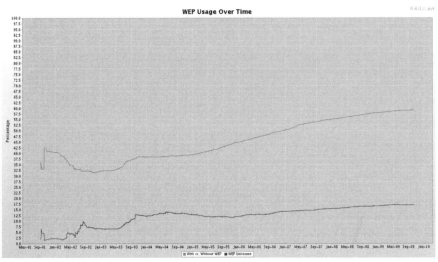

FIGURE 1.7

Wigle.net Graph of Encryption Usage

Note the bump in the curve starting just before the marker for May 2007. It is pretty telling that people started paying more attention right after the TJX breach announcement.

Since the disclosure of the TJX penetration, corporate and individual attention to wireless security has increased for the better. Companies and individuals not wanting to be the next poster child for wireless security took efforts to upgrade from WEP to WPA and WPA2 security. While this is a definite step up in security, it is not without its issues.

Cryptographically speaking, WPA and WPA2 solve a lot of the issues of WEP. The IV counter is now larger (48 bits) and various countermeasures are in place to prevent the problems that plagued WEP. Further improvements include replacing the simple cyclic redundancy check (CRC) on each packet with the message integrity check (MIC), or Michael checksum and packet sequence enforcement, to prevent ARP replays and similar attacks.

There is one problem that designers have so far been unable to engineer away. A problem that has been plaguing computer security since the beginning: human stupidity. People are just not very good at choosing hard-to-crack passphrases.

WPA AND WPA2 IN A NUTSHELL

WPA version 1 was an interim solution created by the IEEE when it was clear WEP was on a path to ruin. Fearing mass obsolescence and the backlash from consumers feeling they were being forced to buy new equipment only a short time after investing in it to achieve security, WPA was designed to operate on the limited hardware resources of APs designed for WEP. To do this, WPA still uses RC4 as its stream cipher which limits the load on the equipment. WPA2 (also known as 802.11i) is the final and more secure version of WPA. WPA2 uses Advanced Encryption Standard (AES) as its stream cipher, which is vastly more secure but requires resources only found on the newer generations of APs and is not available on older equipment.

> **NOTE**
>
> Both versions of WPA can operate in two modes. These are Pre-Shared Key (PSK) and Enterprise or RADIUS mode. There are different settings within each mode, but these are the major functions.
>
> In PSK mode, the AP itself handles the authentication and contains a secret key that both the client STA and AP use to set up a secure connection. In Enterprise mode, a connecting client has all traffic blocked except to the authentication server. There the AP can pass further credentials such as usernames and passwords or certificates, which can be integrated with a larger authentication scheme such as Active Directory.
>
> For the most part, PSK is the most common solution for home and small businesses. Enterprise mode requires further infrastructure and more complexity but has the added benefit of central control and integration with directory services for things like single sign on.

In the process, the designers of WPA and WPA2 set about to solve other problems besides security. One of the problems with WEP was the keys themselves, though, not from a security standpoint but from a standpoint of users getting annoyed and turning off security.

WEP keys are usually 26-digit hexadecimal numbers (shorter if you are using less than 128 bits) that had to be typed into the client manager software to connect. These keys were long, complicated, and easy to mistype causing no end of headaches trying to get a system connected. In addition, they were very difficult to remember and thus ended up almost always being written on a sticky note somewhere near the AP or the users' desk. To make things worse, most operating systems (OSs) made you type it manually twice, while masking what you typed behind stars or black circles.

To make things worse, there was no standard definition as to how keys should be entered. Some equipment defaulted to using the key in hexadecimal; others used it in ASCII format. It was not always apparent which format was needed. This caused no end of problems, particularly for people using different manufacturer's APs and client adapters. Often someone would buy an Apple Airport, which used an ASCII key, and would attempt to use their non-Apple computer with its built in wireless to connect. The other OS would request the key in hexadecimal and the AP would expect an ASCII key causing the connection to fail. More often, this leads many users to set the AP to the last known working conditions: no security.

To solve this annoyance, the IEEE included detailed guidance in the specifications on generating the hexadecimal key from an easier to remember passphrase. Both WPA version 1 and version 2 in PSK mode use 256-bit hexadecimal keys (as opposed to WEPs 128-bit maximum) generated with the PBKDF2 algorithm.

The PBKDF2 algorithm generates a key called the Pairwise Master Key (PMK) that is then used to drive a four-way handshake to authenticate client devices to the network.

This algorithm basically works like this:

PMK = (Passphrase + SSID + SSID Length),* 4096 SHA1 iterations

The PMK is made up of the user supplied passphrase (from 8 to 63 characters), concatenated with the SSID of the network as a salt value and the SSID length. This is then run through 4096 iterations of SHA1 and outputs a 256-bit value, which is the key used in the handshake.

The nice part about this system is that all a user has to remember is a simple passphrase – the rest is generated for them since the SSID is known and so is the length. No more having to remember long and ugly hexadecimal strings. You can sit down at any WPA client manager and generate the same hexadecimal key if you know the passphrase and SSID.

Around the time WPA version 1 came out, the security community took note of an interesting statement in the specifications:

> *A passphrase typically has about 2.5 bits of security per character, so the passphrase mapping converts an n octet password into a key with about 2.5n + 12 bits of security. Hence, it provides a relatively low level of security, with keys generated from short passwords subject to dictionary attack. Use of the key hash is recommended only where it is impractical to make use of a stronger form of user authentication. A key generated from a passphrase of less than about 20 characters is unlikely to deter attacks.*[2]

This statement, from the 802.11i specifications, indicates that any passphrase of fewer than 20 characters was considered weak, as anything less was not random enough and could be subject to a dictionary attack. This led to Robert Moskowitz writing an article for wifinetnews.com that explained that since an attacker knew the SSID of the network they were attacking and the SSID length, they had several of the parts that are used to make up the key and that the entire security of the key rested on good passphrase selection. Not a good prospect from the security community view of things.

The article also pointed out that nothing was secret about how keys were generated and that the hashed version of the key is sent between the STA and AP and can be easily captured. In theory, an attacker could run a dictionary brute force attack where they insert their dictionary word into the PBKDF2 algorithm and run it through the 4096 SHA1 iterations. The resulting value is compared to the captured hash, and if they match, then the attack knows the plaintext key.

The limitation is that there are a staggeringly large number of possible passphrases. Passphrases can be between 8 and 63 characters long. There are 94 possible characters (ASCII characters 32 to 126) that can be used for each character of a passphrase – this includes all the upper- and lowercase letters, numbers, and symbols on a keyboard, leading to a very large number of possibilities. SHA1 is also very CPU intensive and takes a comparatively long time to calculate, so to do an exhaustive search would take thousands of years or longer.

Where the weakness enters the picture is the one element of the formula that is not in the math; the human operator. Since the key is up to the user to select, they will often opt for an easy way to remember passphrase based on dictionary words. This greatly reduces the key space needing to be searched by an attacker. If they know to limit their search to dictionary words, they can limit the search to a few hundred thousand words instead of the trillions of possible combinations. Where the real rub occurs is that the attack only requires that the attacker can obtain the four-way handshake and they can spend all the time in the world to run through world lists offline and away from the target.

This weakness was first implemented by Josh Wright in his tool coWPAtty. Provide a list of dictionary words, the SSID of the network you are attacking, and a capture file containing a hashed version of the key, and coWPAtty will run through the dictionary and hash all the words out and see if they match the capture. If they do, you have the passphrase.

First versions of the program were limited to running the attack straight through, meaning that the hashes had to be calculated each time it was run. This was annoying if you were testing the key on multiple networks with the same SSID. The CPU power was being wasted since it was being repeated multiple times.

Shortly after the release of coWPAtty version 2, a wireless research group, the Church of Wi-Fi, undertook a project to speed up this process.

NOTE

In the world of password hash breaking, there is a concept of a 'rainbow table.' This is a situation where instead of attacking each password individually each time and starting the process and it's resulting CPU cost each time, you simply apply the concept of a time/space trade-off and calculate the hash for every possible character combination and store the results for later lookup. In 2005, the Shmoo Group (www.shmoo.com) released their rainbow tables to the public that contained every possible password for the LanMan hashes used to store older versions of windows login passwords.

The time/space trade-off is harder to apply to WPA passphrases. Since the SSID and the SSID length are seeded into the passphrase hash, this means that the passphrase of "password" will be hashed differently on a network with the SSID of "linksys" then it will on a network with the SSID of "default." So in order to generate an exhaustive hash list of all passphrases possible for all networks, one also has to do so for each SSID possible, which is obviously a lot and would take a huge amount of storage, literally thousands of terabytes, if not more.

The Church of Wi-Fi took a different approach to speed up the process. They figured that if you run the attack against an SSID once, why not store the list of resulting test hashes to use again later if the same SSID was encountered. For example, if you come across the SSID "wireless" and run a dictionary against the key, even if you are unsuccessful, save the output, and the next time if you encounter that SSID, you don't have to spend the CPU time redoing the calculations for the same dictionary – just look up the previous output.

This concept was further extended to the idea that if you know the SSID of your target (that is, you are doing a penetration test), you can save your time on site by spending days, weeks, or months of time generating a hash lookup table for that SSID to be used on site. You could set up a few spare machines to work 24/7 before the scheduled test to give you a time advantage on site. In addition, you could save those tables for the next test as well. The result was the addition of the genpmk program to coWPAtty to generate hash lookup tables without the need for a captured handshake.

To demonstrate this improvement, the Church of Wi-Fi set about to make a set of tables that would give the greatest chance of recovering the key, rather than trying to build an exhaustive list of all possible passphrases for all possible SSIDs, which would be absolutely prohibitive to store or calculate in a reasonable time. Their approach was to apply some psychology to the process to build a targeted list. You see, people are predictably lazy and tend to choose easy-to-remember passphrases. In addition, people also choose fairly simple SSIDs or just leave them at the manufacturers default.

Taking lists of common and known-used passphrases, along with wigle.net's list of the top 1,000 SSIDs (which accounts for approximately 50 percent of their database of tens of millions of networks), they computed tables consisting of approximately 1.2 million common words and passphrases for each of the top 1,000 SSIDs as a proof of concept. These tables can be used if you are attacking or auditing (arguably the same thing with different intentions) a network configured with one of the top SSIDs – you can quickly do a lookup and compare the pre-hashed tables to your captured hash several orders of magnitude faster than doing the CPU calculations onsite. Tests on a Pentium 3,700 MHz laptop showed at the time it could test 12 keys/s whereas to just do a lookup from a table it could test 18,000 keys/s.

The Church of Wi-Fi tables are distributed free via bittorrent links on churchofwifi.org or by DVD sales from www.renderlab.net/projects/WPA-tables for the bandwidth impaired.

Furthering the idea behind this project, the folks at offensive-security.com (the group behind the Backtrack bootable Linux security distro) generated 500 Gb of tables using a similar methodology. While they eliminated some of the more extraneous SSIDs from the Church of Wi-Fi list, they cover all the very common ones. Each SSID is 1.9 Gb and uses a 49 million word dictionary. Obviously, this is a bit hard to distribute, but they are available via bittorrent at www.offensive-security.com/wpa-tables/

HOW TO CRACK WPA PSK AND WPA2 PSK

Since WPA and WPA2 share the same key generation mechanism, the same attack on password selection works on both. When coWPAtty was first altered to add support for WPA2, it was only a few lines of code in the parser that needed to be changed; the rest of the code was fine as it was.

The ability to crack a WPA key is based on two things: the quality and size of the dictionary used and the amount of time an attacker is willing to invest. If the passphrase used is not in the dictionary supplied to coWPAtty, there is no chance of recovering the key. However, if you use a huge dictionary or try to do an exhaustive search, you'll be sitting there somewhere on this side of forever waiting. Depending on your intentions, on a professional audit it may be easier and more feasible to simply ask for the passphrase and ensure it would not be in a dictionary likely to be used by an attacker. In the situation of a penetration test or an actual attack, manually testing the hashes is the next option.

Your best expectation is to audit a network and make sure that the password is not (or is, depending on your intentions) in any reasonable-size dictionary available to an attacker.

The first step is to capture the four-way handshake of a client authenticating to the network. Every client does this and in the process, a hashed version of the key is sent through the air. Once we have that, we can attack it.

How you capture the handshake is up to you. Packet capture tools like Wireshark (www.wireshark.org) can be used, but the Aircrack-ng suite's Airodump-ng provides probably the simplest interface. See the WEP cracking tutorial in this chapter on setting up Airodump-ng to listen to a target:

```
airodump-ng --channel 6 --bssid 00:16:b6:1c:91:94 --write bar ath1
```

Airodump-ng, in addition to collecting packets for WEP cracking, also can alert when a four-way handshake has been captured too. When a valid handshake is captured, in the upper right-hand corner of the Airodump screen will appear a "WPA Handshake" along with the BSSID of the network it caught, as in Figure 1.8.

It may be a while before a valid client joins the network, and even then the full handshake may not be exchanged or captured. It may be necessary to force the issue with a deauthentication attack.

A deauthentication attack (deauth) is where an attacker fakes a message from the AP to the client asking it to disconnect. Normally, this is used to gracefully close sessions if the AP was going down or rebooting. Since these sorts of administrative messages are unencrypted and unauthenticated, anyone can inject them and the client will obey and disconnect. A few seconds later the client sees that the network is back (it never left) and then reconnects, thus disclosing the four-way handshake.

Aireplay-ng has an option to perform this attack, along with other tools, but aireplay has the added advantage of requiring a specific target and avoids collateral damage from unintended disconnects of other networks:

```
aireplay-ng -0 5 -a 00:16:b6:1c:91:94 -c 00:0e:35:76:53:47 ath1
```

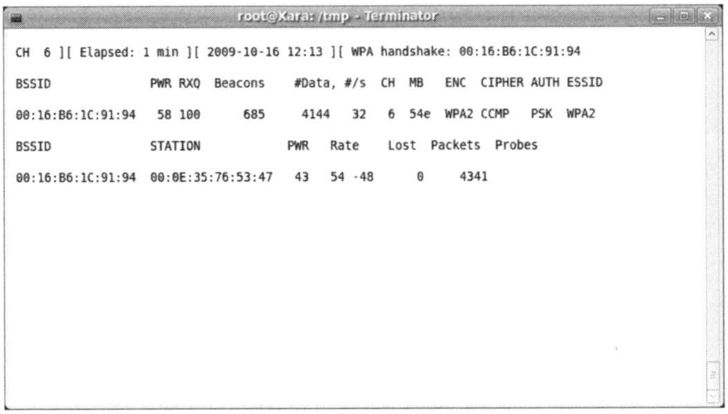

```
root@Xara: /tmp - Terminator

CH  6 ][ Elapsed: 1 min ][ 2009-10-16 12:13 ][ WPA handshake: 00:16:B6:1C:91:94

BSSID              PWR RXQ  Beacons     #Data, #/s  CH  MB   ENC  CIPHER AUTH ESSID

00:16:B6:1C:91:94  58 100    685        4144   32   6  54e  WPA2 CCMP   PSK  WPA2

BSSID              STATION            PWR  Rate   Lost  Packets  Probes

00:16:B6:1C:91:94  00:0E:35:76:53:47   43   54 -48    0     4341
```

FIGURE 1.8

Airodump-ng Captured a Handshake

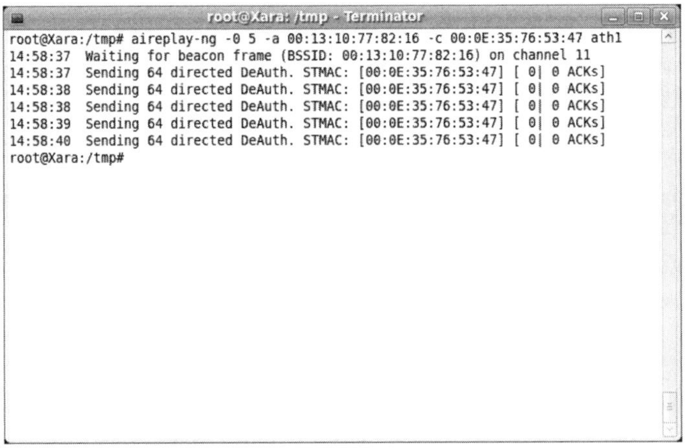

```
root@Xara: /tmp - Terminator
root@Xara:/tmp# aireplay-ng -0 5 -a 00:13:10:77:82:16 -c 00:0E:35:76:53:47 ath1
14:58:37  Waiting for beacon frame (BSSID: 00:13:10:77:82:16) on channel 11
14:58:37  Sending 64 directed DeAuth. STMAC: [00:0E:35:76:53:47] [ 0| 0 ACKs]
14:58:38  Sending 64 directed DeAuth. STMAC: [00:0E:35:76:53:47] [ 0| 0 ACKs]
14:58:38  Sending 64 directed DeAuth. STMAC: [00:0E:35:76:53:47] [ 0| 0 ACKs]
14:58:39  Sending 64 directed DeAuth. STMAC: [00:0E:35:76:53:47] [ 0| 0 ACKs]
14:58:40  Sending 64 directed DeAuth. STMAC: [00:0E:35:76:53:47] [ 0| 0 ACKs]
root@Xara:/tmp#
```

FIGURE 1.9

Aireplay-ng Running a Deauthentication Attack

In this case, we are using option "0" to specify the deauth attack. Five is the number of times to do this attack, just to make sure the client disconnects. The target network BSSID is specified, as well as the client MAC and the injection interface. Once this command runs, Aireplay-ng will deauth the client five times, and hopefully when the client reconnects, Airodump-ng will capture the handshake. Aireplay displays output similar to Figure 1.9, substituting your target addresses, of course.

Once you have a valid four-way handshake, you can start the cracking process. The nice part of this attack is that all it requires is the handshake, which can be captured passively in a few seconds if you are lucky. The rest of the attack can take place offline elsewhere and does not require the target network.

TIP

Capturing a complete 4-way hash is trickier than it sounds. Differences in client managers, AP behaviors, and even being on the edge of range can lead to incomplete captures. If you are having trouble in capturing a complete handshake (if coWPAtty is reporting so), try adjusting your location relative to the AP and the client and try again. All else fails, wait for a client to log in normally as some clients behave oddly on reassociation after being deauthenticated.

CoWPAtty only needs a few pieces of information to begin the attack once the data has been collected – in this case, the wpa2psk-linksys.dump capture file:

```
cowpatty -r wpa2psk-linksys.dump -f dict -s linksys
```

In this case the capture file, which is wpa2psk-linksys.dump, the dictionary file, dict, and the SSID of the network we are attacking, in this case, linksys. CoWPAtty will parse the capture file, and if there is a complete handshake, the crack will begin; otherwise, it will report that the handshake is incomplete and you should try to reacquire it. If you have a complete handshake, coWPAtty sets about computing each word in the dictionary through the PBKDF2 algorithm with the specified SSID and comparing the output to the capture. If they match, it reports the successful passphrase. If not, it moves on to the next one. Figure 1.10 shows that the PSK on this network was dictionary.

Depending on the CPU you are using and the size of the dictionary, the length of time to run through the list will vary. If there is a match, however, coWPAtty

FIGURE 1.10

CoWPAtty Successfully Retrieves the Passphrase

will report the success. In Figure 1.10, the CPU could work through approximately 100 passphrases per second and went through the fairly short dictionary (approximately 10,000 words) in 37 s. If this is an SSID that you audit regularly, you may not want to spend all the CPU time each time you want to run through this same list. In that case, you will want to save the output in a lookup table.

EPIC FAIL

CoWPAtty expects the dictionary to be a simple list of words in UNIX test file format as opposed to DOS formatted. DOS-formatted text has a hidden control character at the end of each word, whereas the UNIX format does not. If you feed coWPAtty a DOS-formatted list, its parser includes the hidden control character as part of the passphrase and is computed as such. So the passphrase of "password" used in the list will actually be computed as "password/r," and of course, the resulting hash will not match the capture for a network using the passphrase "password." The Church of Wi-Fi had this happen to them twice in the course of generating their tables, and they had to throw out months of work.

The mathematics behind the lookup table is exactly the same as running a "live" crack, except instead of comparing to a capture file, the hashed version of the word and its plaintext version are written to a file. That file can then be looked at later to quickly look for matches. To generate your own lookup table, coWPAtty comes up with the program, genpmk:

```
genpmk -f dict -s linksys -d linksys.hash
```

Genpmk takes a word list (dict), network SSID (linksys), and outputs the resulting hashes to a file (linksys.hash). This process takes just about as long as running coWPAtty directly on the capture as the time taken between Figure 1.10 and Figure 1.11; however, this file can then be fed back into coWPAtty for much faster lookups on subsequent tests.

The most likely times you will want to compute a hash file is if you know the target ahead of time and want to spend off hours (night time or other idle CPU time) to give yourself a time advantage onsite (spend the time now to save it onsite) or if you repeat the same operations over and over and want to save the time by investing it only once and saving the output for later use. If it takes you an hour to compute through a word list, rather than spend that hour each visit, if you do it once and save the output, you can effectively save an hour each time.

Once you have a hash file generated, you can enter it in place of, or in addition to, a dictionary. CoWPAtty will quickly look through the hash file for matches. If none is found, it will report the failure, or if you specified a word list, it will begin the CPU intensive crack with that list. If you look at the speed of the crack between Figure 1.10 and Figure 1.12, you will see the marked improvement in speed:

```
cowpatty -r wpa2psk-linksys.dump -d linksys.hash -s Linksys
```

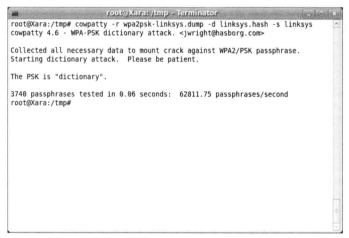

```
root@Xara: /tmp - Terminator
root@Xara:/tmp# genpmk -f dict -s linksys -d linksys.hash
genpmk 1.1 - WPA-PSK precomputation attack. <jwright@hasborg.com>
File linksys.hash does not exist, creating.
key no. 1000: apportion
key no. 2000: cantabile
key no. 3000: contract
key no. 4000: divisive

4090 passphrases tested in 40.37 seconds:  101.30 passphrases/second
root@Xara:/tmp#
```

FIGURE 1.11

Genpmk Computing a Hash File

```
root@Xara: /tmp - Terminator
root@Xara:/tmp# cowpatty -r wpa2psk-linksys.dump -d linksys.hash -s linksys
cowpatty 4.6 - WPA-PSK dictionary attack. <jwright@hasborg.com>

Collected all necessary data to mount crack against WPA2/PSK passphrase.
Starting dictionary attack.  Please be patient.

The PSK is "dictionary".

3740 passphrases tested in 0.06 seconds:  62811.75 passphrases/second
root@Xara:/tmp#
```

FIGURE 1.12

CoWPAtty Using a Precomputed Hash File Instead of a Dictionary

In Figure 1.12, doing a live crack onsite would have been done in about 37 s. Using a hash file, which could be computed ahead of time in off hours or down time or from one of the publicly available sets, the lookups were done in 0.06 s. Quite the time savings, if you end up reusing the hash file several times.

Recent developments have helped improve the process of doing the computationally heavy portions of WPA cracking and hash table generation. The usage of

Field Programmable Gate Arrays (FPGAs), which are programmable processors, has shown a much faster SHA1 implementation than on generic CPUs since the algorithm is built into the hardware itself. This means that the FPGA does only one thing and it does very fast. Additionally, recent advances in video card Graphics Processing Units (GPUs) from companies like Nvidia and ATI have allowed non-video processes to be run on their chips. These chips are much better suited for the type of computations being performed. A sample implementation is available at http://code.google.com/p/pyrit/

SUMMARY

While WPA-PSK (and WPA2-PSK) is an improvement over WEP, by creating a simpler method for users to remember and enter passphrases, the IEEE introduced a human flaw into the equation. WPA-PSK has a 256-bit key, much more substantial than WEPs 128-bit offering, and the randomness of that key is based on a user's ability to choose a random passphrase and a unique SSID, which in general, people are not very good at.

In short, it's all down to choosing a good passphrase. There have been no reports of WPA-PSK passphrases being cracked with such an attack; however, there's really no way to detect it being done. Anecdotally from professional pen testers though, usage of the Church of Wi-Fi tables has enjoyed about a 50 percent success rate if the SSID is in the list, meaning that an attacker has a decent chance of being able to walk into a situation and being able to crack a password. It just goes to show that if you use WPA-PSK, you had better follow the IEEE's advice from the standard and use a 20-character passphrase that is not based on any dictionary words, and is made up of uppercase and lowercase letters, numbers, and a few symbols for good measure.

Other measures to reduce your risk of a brute force attack on WPA:

- Periodically changing passphrases (maintaining length and randomness)
- Periodically changing the SSID, thus changing the salt value
- If sensitive or financial data is being sent, consider the installation and use of a Wireless Intrusion Detection System (WIDS)
- If at all possible, migrate to a WPA-Enterprise solution
- Educate your users on the need for complex passphrases

Endnotes

1. www.netstumbler.org/f22/faq-legalities-concerning-wardriving-netstumbling-nethugging-6430/ #post48649; [retrieved 01.12.09].
2. http://standards.ieee.org/getieee802/download/802.11i-2004.pdf; [retrieved 01.12.09].

Wireless – Client Attacks

Wireless network security is getting better. When wardriving first hit in 2001, statistics showed only about 30% of access points were secured with some sort of encryption. Today that number is close to 70%, indicating that people are slowly getting better at securing their access points. Since then, wireless client devices have become a tempting target for attackers. Workers become increasingly mobile and depend on wireless network access at the office and on the road. The availability of public hotspots has the general public becoming more and more accustomed to Internet access being available at every coffee shop, airport, and hotel they might travel to.

Client security is often overlooked as users often think of it when connecting to the "home" network. In reality, clients are sometimes more of a tempting target since they can be attacked away from watchful eyes of information technology (IT) departments and often can contain just as much valuable information or be useful entry points into the network.

In early 2006, Simple Nomad of the Nomad Mobile Research Center (NMRC) presented a talk at Shmoocon in Washington, D.C., outlining his experiences in a talk entitled "Hacking the Friendly Skies." In his talk, he detailed the many ways he was able to access the private documents and manipulate the information on laptops in use at 30,000 ft on cross-country flights. His discoveries and techniques showed that he was able to find at least one vulnerable target on every flight he tested on, often without any indication or interaction with the user. His actions showed that every executive in the first class with a laptop is a potential target, even in more unusual environments, far away from the watchful eye of IT administrators (www.nmrc.org/pub/present/shmoocon-2006-sn.ppt).

Vacationers at hotels are often a tempting target. There is no guarantee that the hotel-named hotspot you are connecting to is actually the hotel's network. Take the case of the Defcon 16 hacker conference at the Riviera Hotel and Casino in Las Vegas in 2008. Before the convention, there were 12 access points detectable from the pool with the hotel's network Service Set Identifier (SSID). After the first day of the three-day conference, there were 130 access points available. Some odds at casinos are more in your favor than others, and a hapless traveler connecting to a legitimate access point is not one of them.

PUBLIC HOTSPOT DANGERS

Just about everyone is familiar with the idea of public Wi-Fi hotspots. Every coffee shop seems to have one, as well as restaurants, bars, grocery stores, airports, and hotels. Hotels, in particular, are nearly required to have wireless Internet access to stay competitive and, with today's seemingly endless flight delays, airport wireless is probably the only thing keeping business travelers from staging a revolt.

What most people don't seem to think about are the inherent dangers of bringing a computer onto a random network that you have no reason or way to verify or trust. Would you haul a desktop computer around and connect it to any random network cable that you found lying around? Of course not (for reasons other than the obvious hauling problems), yet people connect their mobile wireless devices to seemingly random networks all the time without thinking about who else might be on the same network.

From an attacker's perspective, the neighborhood coffee shop where all the customers hang out on sofas with laptops is a prime ground for targeted fraud and abuse of wireless clients. Hotels hosting conferences are also a very tempting target.

HOW HOTSPOTS WORK

There is no set standard for public hotspots. They are designed around a business model rather than a technical model. Their main purpose is to draw in and get customers to stay long enough to order a second (or third) cup of coffee.

In the early days of wireless, it was as simple as the owner running a digital subscriber line (DSL) or other high-speed line into the business and plugging in an open access point. This worked great: customers could just connect with no issues. Shortly thereafter the inevitable happens and all the neighboring homes and businesses notice the new open network and proceed to use it freely without purchasing anything from the owner. After a short time, legitimate users complain about the slow speed, and the owner figures out that there are a few dozen extra users on the system than in the business. They then proceed to enable encryption and put up a sign for customers to ask for the password to use the wireless. This works for a while until the neighbors figure out that they can go in and buy the cheapest thing on the menu and get the network key and proceed to use it as before. The owner again figures out why the system has suddenly slowed and starts a regime of changing the key each day. After a while, this gets cumbersome and more trouble than it's worth, which leads the owner to finally disable it for everyone.

This is a classic example of the tragedy of the commons, a game theory[A] concept where one person taking more than his or her share of a common resource is not an issue, except that everyone thinks the same way and the group ends up over-consuming that resource, thus destroying it for everyone. The common example is a group of sheepherders leading the herd to grazing on public land. If one adds an extra few sheep, the additional grazing won't be that conspicuous. The problem arises from everyone thinking the same way and expanding their flocks. Soon overgrazing means none of their larger flocks get enough food and therefore the sheepherders won't profit.

The eventual solution to this in the wireless world was the captive portal and the pay-for-play model. This system is what you will find most often now and has been turned into a fairly sizable industry of providing turnkey solutions to our hapless coffee shop owner and his ilk.

Captive portals typically operate as open networks with one difference. Once connected to these systems, they block all traffic and redirect all Web requests to a local page. That page is usually an authentication page that explains how to access the Internet at large. Some of these only require acceptance of a term of service to give owners some legal recourse against abusers. Others go a step further and require payment for a set amount of timed access. Once payment is made, that particular computer is allowed to communicate with the outside world for a prescribed period of time.

This system grew into companies operating large networks of access points around the world sharing a common authentication and payment scheme where you can purchase a set period of time, say 30 days, and use any of their access points around the world. This is typically the system you would have at an airport or hotel, since on a long journey you might need Internet access at many different points during your travels.

The growth in use of captive portal and pay-for-play configurations is when attackers first started to take notice of these systems (as a target for fraud). Since most payments were made via credit card, the opportunity was present to capture those card numbers and other interesting information.

ATTACKING PUBLIC HOTSPOTS

Most public hotspots, be they free or for pay, generally are not encrypted. A chicken and egg situation exists in how to give the customer an encryption key from an automated system without also providing it to noncustomers. Employees should be

---

[A]For more information on game theory, the Web site www.gametheory.net/ is a good place to start. You will never look at games the same way again.

making coffee and not doing technical support. If the network is encrypted, then users can't get a key, and if they can't get a key, they can't connect. As a result, most common public hotspots are unencrypted.

NOTE

While they are unencrypted, most often the login page or payment page is secure sockets layer (SSL) encrypted. This means that under normal operation, credit card numbers are not being sent in the clear. If the pages were not encrypted, the attacker's job would be instantly simple. An attacker would just have to use a wireless sniffer utility like Kismet (www.kismetwireless.net) or Wireshark (www.wireshark.org) to capture the traffic and then set about looking through the packet dumps for those page requests and the resulting form submission.

As an attacker, having access points unencrypted gives an extreme advantage. The user has no way of actually verifying that the access point is what it says it is since there's no shared secret. A network's SSID is merely for organizational purposes, it is not an authentication method. Just because you are in a Starbucks and your laptop shows a network in the area named *Starbucks* does not mean it is owned nor has anything to do with that Starbucks (though it might be from the one across the street or further down the block). Many people assume that the SSID is somehow an indication of authenticity. It is far from that. As a simple test, set your home access point to "Starbucks" or "tmobile" or any other pay-for-play hotspot name. There is nothing stopping you.

Many wireless cards on the market are capable of operating in "master" mode, which allows them to operate as an access point. This fact, along with people's trusting nature, allowed early attackers to impersonate access points easily. Originally, it would take a fair bit of work to set up but later advances would make it simpler to execute this type of attack. Most laptops were not designed or intended to be used as access points; however, the capability existed since the same card that powers the Wi-Fi in your laptop can also run an access point.

The attack would work like this. An attacker would set up his or her laptop or wireless device in master mode and become an access point. The SSID would be set to either impersonate an existing access point or appear to be a legitimate access point where there was none. In the case of impersonation of an existing access point, clients will generally connect to the strongest signal if several of the same SSID is seen, so an attacker just has to be closer or have a more powerful signal than the legitimate access point. (An alternate method would also include a denial of service attack on the real Access Point [AP].) From there, the attacker can set up Dynamic Host Configuration Protocol (DHCP), domain name system (DNS), and a Web server to emulate the login page of the legitimate captive portal and present that to users as if it were the legitimate access point. The attacker has control of all network services needed to "surf" and is able to harvest usernames, passwords, and credit card numbers. If the attacker wanted to further cloak his or her existence, the addition of a second wireless card connected to the legitimate access point could pass the victims' information and other traffic to its intended destination, further hiding the man in the middle (MitM) (that is, the victim is able to access the Internet and receives no indication that an attacker is "listening in"). This type of attack opens up other opportunities for data harvesting and abuse, discussed later in this book.

In 2003, to bring some attention to this, the Shmoo group released "Airsnarf" (http://airsnarf.shmoo.com), a series of scripts and a tutorial on how to set up a system to achieve the above scenario and harvest usernames and passwords from a fake captive portal. This version of the attack was somewhat cumbersome and hard to pull off surreptitiously. However, the idea was there. The down side was that the software could only impersonate one access point at a time and the attacker had to wait for clients to connect. There could be times where no one was using the network for days and the effort is wasted. There had to be better (or worse, depending on which side you were on) methods to get clients to connect to the fake network.

Around the same time as Airsnarf, another program, "Hotspotter" was released that partially automated some of the process of luring clients.

Hotspotter was written in response to an interesting behavior of Windows XP (preservice pack 1) where wireless clients could be coaxed from a secure network to an insecure network by an attacker simply setting up a network with the same SSID that was unencrypted and a stronger signal than the secure network, and then sending a spoofed deauth command to the client. The client would reassociate to the strongest signal (the unencrypted, rogue network) and not pop up any alerts to the user that anything had happened or that they had changed to an insecure network. This pop-up behavior was fixed in SP1 for XP, and raised awareness, both good and bad, that some client managers have interesting and potentially dangerous behaviors in how they deal with storing and reconnecting to preferred networks.

The behavior being exploited is one of convenience. Users will often connect to many different networks (work, home, coffee shop, airport, and so on) and each will require different settings. Most, it not all client manager software will remember networks (usually SSID, network keys, and so on) that it has connected to before and store those settings. In some client managers, the default behavior when not connected to a network is to automatically connect to these networks it has connected to before (in other client managers, there is a setting to enable it to automatically connect; otherwise the system asks first). This causes an interesting situation when you have a common SSID setup in your preferred network list.

"Linksys" is the most common SSID in the world according to wigle.net,[1] with 8 to 9% of the wigle.net database at any given time. These networks are using the default SSID for linksys wireless devices and often if the device is using the default SSID, the rest of the device is in its default state, namely unencrypted. If a user has "linksys" entered in his or her preferred network list, many systems only distinguishe networks based on SSID and will assume any network named "linksys" is the original one entered in the list and attempt to connect. This was often known as *network drift* in the wireless community. Many client managers would prefer the network with the highest signal strength, and there was a good chance that there was an open linksys access point around and clients would suddenly find themselves connected to a different network than they intended.

NOTE

The SSID is what most people call the Extended Service Set Identifier or ESSID. It is the logical name of the network as seen by users. The Basic Service Set Identifier or BSSID is typically referred to as the MAC address since it is the same as the MAC address for the wireless access point.

Hotspotter listens on a wireless interface in master mode. When it hears probe requests from a client, it checks a list of common commercial hotspot names and if it finds a match, it configures itself as an access point being sought, and the client (often) will automatically connect. From there, it is up to the attacker to have Hotspotter trigger a DHCP daemon, port scan – whatever he or she wants to do.

Again, this behavior could only emulate one target at a time. If you were attacking a large group, such as you would find at a conference, then you would want to be able to configure for multiple-probed SSIDs, and not just a predetermined list.

In 2004, Dino Dai Zovi and Shane Macaulay wrote and distributed a program called *Karma*. This software was what attackers had been seeking for a while. Karma operated similar to Hotspotter in listening to probe broadcasts from clients and replying. The difference, however, was their patches to the Madwifi driver used under Linux that allowed the dynamic creation of replies to probes on the fly. This meant that if a dozen clients were all probing for different networks, Karma would be able to dynamically reply to those probes and association requests for any SSID.

The conversation simply became the following:

Client 1: "I'm looking for mycorpnet. Are you mycorpnet?"
Karma: "Sure, I'm mycorpnet, come on in."
Client 2: "Are you supersecretninjanetwork? I am looking for supersecretninja-network."
Karma: "Why not? I'm supersecretninjanetwork too, connect to me!"
and on, and on, and on....

Karma also came along with a suite of tools to emulate common services. If the connected client had an e-mail client running, it would probably try and connect when the network connection was established. Karma would see the pop3 request and record the login username and password. Without a second card providing a back haul to the Internet, the request won't go through but now the attacker has your password and server address. Same goes for DNS requests being redirected, Web page requests, and other common services. Quite simply, the attacker has established himself as an intermediary (also known as MitM) and can see and record everything sent from multiple clients, while these clients think they are talking to different (authentic) networks.

TIP

The success of this attack depends on many factors, including signal strength and being a more tempting network to connect to. Try a better or more directional antenna or a higher power network card, or simply move closer to the target. Often you will find yourself competing with the access point in the room and you must be able to overpower it.

Also, some recent service packs and updates for various operating systems require user input or some sort of interaction before connecting to an open access point, even if it is in the preferred list. That said, there are still some client managers and other devices that automatically require their network connections when in range.

In response to this new level of attack, wireless client vendors tightened up their software and now many will not connect automatically unless specifically selected (that is, the user has to "click to connect"). That said, the software will still populate a list of nearby networks to select from, including those that were probed for and Karma replied to, so unwary users could see a common commercial hotspot and think it's alright to connect, only to find themselves connected to an attacker now monitoring all their communications. Even if the victim realized right away that there was something amiss and disconnected, the damage is probably already done. Many applications today, from Web browsers to e-mail clients, Really Simple Syndication (RSS) readers, and so on will automatically try to connect when the network is connected, so as soon as the victim connects, these applications automatically reach out and try to connect, often divulging usernames, passwords, and other information in a matter of seconds.

If it didn't seem dangerous enough, in 2008 Karma was integrated into the Metasploit framework. The Metasploit framework is a penetration testing tool to help build and execute exploits as well as including a library of several hundred exploits ready to go. The framework makes this easy by taking care of a lot of the normal overhead of delivery mechanisms, payloads, and so on. At the time of writing, it had not been fully integrated and required some external software to run (a DHCP server to hand out addresses and Airbase-ng to respond to probes) but was at a stage of development where client probes were responded to and clients could connect, insecure protocol credentials would be logged by the framework, and if the attacker so desired, an attack could be injected into any legitimate-requests Web pages. Further integration is expected and should overcome the complication of requiring external software, improving the ability to automatically scan and exploit any client that connects, thus making this a truly terrifying tool in the hands of an attacker, but a very useful tool in the hands of a penetration tester.

> **NOTE**
>
> Although improvements to clients, such as Windows Zero Config, have made it slightly harder to get completely automated connections from modern Windows operating systems (OSes), there are a lot of other devices still very vulnerable to this kind of situation. Cell phones now feature Wi-Fi capabilities, smart phones, MP3 players – even things as digital picture frames and game consoles use Wi-Fi and can be coaxed to connect to an attacker. Threats are not always apparent.

Attackers consider one of the best environments for these sorts of attacks to be 6 miles in the air. Who has not been on an airplane, particularly long-haul flights to business centers, such as London or New York, where as soon as the seat belt sign is turned off, laptops come out and everyone from corporate executive types to vacationing surfer dudes pull out their laptops to do some work, watch a movie, or play games? More often than not, they ignore the flight staff and Federal Aviation Administration (FAA) warnings about wireless devices in flight and do not disable their Wi-Fi adapters (which still drain power from limited batteries) and they continue to probe for networks.

> **WARNING**
>
> Unless the flight crew specifically allows wireless (that is, the flight has in-flight Wi-Fi), it is generally forbidden by many countries' federal laws to use Wi-Fi or any wireless devices in flight. Keep that in mind. However, the chances of Wi-Fi interfering with flight systems are terribly small. If it was a true danger, in-flight Wi-Fi would not be allowed and laptops in general would be banned, since planes falling out of the sky because passengers booted Windows and forgot to disable their Wi-Fi cards is not a good situation to have. In general, the regulations are in place as a precautionary measure and are a good idea because 6 miles above the earth and traveling 500 miles an hour is not a normal situation for humans, and gravity is a harsh mistress.

While a few airlines offer in-flight Wi-Fi service (those are ripe for a whole other set of attacks), most do not. The environment of a plane is perfect for attacking wireless. Sounds strange, but consider that there are no other networks around, so if you bring up your own using Karma or a similar tool, you will be the only network they see. Being in a long metal tube over an ocean also limits any interference, so you'll often find that from cattle class at the back of the plane, you can catch probing clients in first class. Once they have connected to you and you as an attacker hand them an address, you now have a direct connection to their computer. You may not be able to provide any Internet access, but you can poke and prod at their system for hours on end without it seeming strange that you have a laptop open as well.

The "Hacking the Friendly Skies" presentations from Simple Nomad and anecdotal evidence from various wireless message boards indicate that a surprisingly large number of devices are not running a software firewall or seemingly any protection from other clients on the local area network (LAN). Open shares with corporate documents, to vulnerable applications ready to be taken over. A well-prepared attacker is in a prime situation to wreak some havoc. Add in-flight Wi-Fi to provide access to vulnerabilities unforeseen ahead of time, and there's no limit to what can happen.

THE CRUX OF THE PROBLEM

In the days of wired networks, many attacks required that the attacker find some way to insinuate himself between the victim and the system he was communicating with in order to monitor. At the very least, attackers had to find a way to get themselves onto the local LAN at one end or the other. Early on with hub-based networks, the attacker could see everything on the wire; later, switched networks required intermediate steps like arp spoofing the gateway for traffic to be rerouted through the attacker. This is generally known as a MitM attack since the attacker is impersonating both participants in the conversation and can monitor or manipulate the conversation as he or she sees fit. With physical constraints, you can limit an attacker's access to the cable network since he or she physically would have to plug in or have remote control over a device in the local LAN.

With wireless, this hurdle is reduced since, as you can see, it is very easy to get clients to route everything through you. With recent vulnerabilities in SSL, even encrypting Web pages or using virtual private networks (VPNs) can be defeated with tools like sslsniff (www.thoughtcrime.org/software/sslsniff/). The worst case scenario with this situation is where users are not intending on using Wi-Fi at all.

Often users will think themselves clever and utilize a non-Wi-Fi connection to connect to the Internet or VPN back into the office. The easy availability of 3G modems and wired network connections at some hotels make this a realistic situation. Although it is a decent solution to the Wi-Fi problem, very often users will not disable their wireless adapter when using another network connection.

Windows and other OSes are often intelligent enough that when multiple network connections are present, only the last, most recently used network will be used or, at least, the last connected gateway. Some systems, for various reasons, have their network interfaces bridged together (such as Windows Internet connection sharing capabilities). This can lead to an interesting situation where the client effectively becomes an access point and gateway to the other network connection. If this includes a VPN connection, the attacker is routed straight through and into the corporate network (or the Internet), creating obvious problems for corporate security.

SOLUTIONS

With tools like Karma capable of impersonating any open access point, it's very difficult to defend against the possibility of connecting to a fake access point. The best defense is to limit what can occur should that situation happen.

The first countermeasure is to use a software firewall with a policy that passes only needed traffic through and denying the rest. Should you find yourself connected to a fake access point, at least you aren't sharing all your files with the attacker.

The next solution is to just avoid Wi-Fi wherever possible and disable your adapter when not in use. If a wired connection is available, choosing that option instead and disabling the wireless adapter while ensuring precautions like firewalls are in place will greatly reduce threats.

Be wary of connecting to anything you cannot verify. When you are at an airport and can see six access points, how do you know if the airport only has five and which is the attacker? Most public hotspots do not publish the BSSID of the access points, so there is no way to know which are legitimate. Those that do publish the SSID on a sign make it easier for attacker to know what name to impersonate. Connecting to public Wi-Fi is a gamble. Even the BSSID for a legitimate AP can be easily impersonated (checking the BSSID does reduce the chance of exploit a bit as it's a step that attackers sometimes do not exploit). Using a 3G or other cellular Internet connection is probably a safer bet.

If possible, install an alternate client manager. Due to standards being what they are, you are not limited to the client management software included with your wireless adapter. Often the manufacturer offerings are simple and cannot enforce any rules about what and where to connect. There are many aftermarket client managers available, typically from wireless security vendors who can provide a bit more logic and sanity to the situation. One example is Airtight Networks SpectaGuard (Secure Agent For Endpoints), which can enforce policies to prevent insecure behaviors, such as connecting to insecure access points (connections must be encrypted), preventing ad hoc connections, and preventing bridging of adapters and only allow connections to specific BSSIDs, providing much more granular selection of access points than just selecting by SSID. This sort of system has obvious benefits for large corporate users who can use existing central management tools to push out such software and prevent users from being able to make dangerous mistakes. Often, such software can also go beyond controlling just Wi-Fi and enforce rules for Bluetooth, WiMax, dial-up, or any other network connections.

INJECTION ATTACKS

In 2004, at Defcon 12, something was strange on the network. Now anyone familiar with the annual Defcon Hacker conference each year in Las Vegas knows that the network setup for that weekend is probably the most hostile network on the planet. You are truly taking your life (or at least your computer's life) into your own hands by connecting since the network is generally known as a free-for-all zone. The common joke is that they should just turn the firewalls around and keep the Defcon network off the Internet: it's easier that way. This year, however, something that was causing confusion was an odd behavior on the wireless network that no one was really able to explain. Given the concentration of intelligence at the conference, this was a rare occasion.

Users typically will sit around the "chill-out" areas and work away on who knows what on their computers. At one point, users started noticing that images on Web pages they were requesting were not what they were expecting. Requests for Google's home page would return Google, but the main image would be different than expected. This same image showed up across multiple pages on multiple sites for multiple users. Of course, being Defcon and the juvenile bent that tends to inhabit attendees, the image was one of a rather horrific nature. Obviously, users expecting Google's iconic logo and receiving a horror show instead led to some truly puzzled users and some amazingly funny reactions (I did mention, "most hostile network on the planet". What were they expecting?)

Various users went about checking DNS settings, checked to see if they were connected to a rogue AP, or went as far as to fire up a packet sniffer to figure out what was happening. Few, if any people (other than the perpetrators) figured out what happened that weekend until the announcement was made later by the perpetrators of this rather hilarious hack. A guy going by the name of Toast took credit and released the tool he coded to do it, called *Airpwn*.

To understand what is being attacked, you must understand how things operate normally. When you request a Web page, your system connects to the remote server with the basic SYN/ACK handshake:

```
Client → Server "GET /foo.html HTTP/1.0" seqno 123 ack 456
Server → Client "<HTML>…" seqno 456 ack 145
```

The client browser sends a GET request for the page specified, "GET /foo.html HTTP/1.0", along with a sequence (SEQNO) number and an acknowledgment number (ACK) for the previous packet. The server normally returns packets containing the requested hypertext markup language (HTML) page, specifying a higher SEQNO and ACK of the previous packet. The process is repeated for every file on the page (images, Java Script, and so on).

Airpwn works off of a simple premise that works like this: On an open network, an attacker can see your packets go through the air much like an old shared medium (hub-based) wired network. A normal Web server's requests will take a bit of time to make the round trip. Since we are in the same room, we can often reply faster than the remote server.

The attack looks like this: A normal request for a file is made, except that the attacker sees it. He then crafts a response with appropriate SEQNOs and ACKs, forging the source MAC address to appear to be from the access point as if it were the legitimate response, except the response contains the attacker's content. Provided the content is small enough to be sent before the real packets arrive, the attacker can close the connection with a Finish (FIN) packet. The original request will now be out of sequence and be discarded. The browser will accept the attacker's content and render it accordingly to whatever damage to delicate sensibilities that may involve. In the sample configurations, there are several different examples. Figure 2.1 shows simple image replacement where images are replaced with the "AIRPWNED" graphic. Figure 2.2 shows replacement of the HTML.

FIGURE 2.1

Airpwn Image Replacement of Google

FIGURE 2.2

Airpwn Replacing HTML

At its core, it's a race. If the attacker can beat the remote server, since the response is expected and its SEQNO is in order, it's blindly accepted as being the legitimate content and passed to the browser.

When it made its debut at Defcon, it was viewed as an interesting party trick and there was some interest in what else could be done with it, but like many tools and attacks, it never seemed to go anywhere. Over time, a few updates were made to adjust to new technology and capabilities, but it became an oddity among the hacker community. It was useful to show content replacement and injection, but not beyond a plaything at conferences to screw with attendees' minds by inserting strange or confusing images into the data in the air.

Airpwn also showed an interesting vulnerability not normally seen. Since it was injecting content that was expected, software firewalls on the client system would happily let the rogue data pass without a second thought. Provided the injected data did not conform to a known virus or detectable attack pattern, it would pass silently through the firewall to the browser, opening many possibilities for abuse.

EPIC FAIL

Airpwn has been a useful tool for driving the point home to users that there are risks in using insecure networks. Explaining the risks only goes so far; injecting alternate images or HTML into the stream and having it show up on their own screen dramatically shows they are at risk. This is effective since it does not do any permanent damage. Those performing these demonstrations need to be careful in choosing the time and the place to perform these demos.

In 2008, at the SecTor (www.sector.ca) conference in Toronto, RenderMan was running a Wi-Fi security clinic, demonstrating various attacks and helping attendees run attacks on demo access points. As part of this, Airpwn was shown injecting data. Unfortunately, it was left running for several hours on the same channel as the conference network. Johnny Long of Hackers for Charity (www.hackersforcharity.org/) came up and asked why the network was misbehaving. Well, it seems Airpwn was injecting extra HTML, preventing Johnny from being able to check in for his flight home. Sheepishly, RenderMan had to terminate Airpwn to let Johnny finish an important task, confirming that there is a time and place for everything, and don't forget what tools you leave running!

EPIC FAIL

Fast forward to 2009 when Dragorn, the author of Kismet, the wireless sniffing tool, and other researchers began to look at what one could actually do with Airpwn and Transmission Control Protocol stream hijacking (as the attack is properly known). This all started with the desire to replace more than images and to have the ability to replace flash video streams on the fly. With the popularity and prevalence of YouTube and other flash-based video sites, it was only natural that they would be a popular target and the desire would bubble up to want to replace any video watched with the "RickRoll" or other suitable content.

NOTE

The RickRoll is an Internet meme where people try and trick others in a bait and switch maneuver into viewing the 1987 Rick Astley music video for "Never Gonna Give You Up." Usually this is through shortened links or redirects, and they are offered to victims as links to other content only to have them end up at the Rick Astley video, and the inevitable joke at their expense by the perpetrator.

   If there is one good thing about RickRolling it's the innocuous nature and generally safe-for-work content. This prank quickly replaced the practice of directing victims to horrific or shocking images and generally will not run afoul of most Web content filters.

Original testing showed that to replace an active stream would be prohibitive as these are generally User Datagram Protocol (connectionless) transmissions. There is no way to silence the stream from the server to insert your own content since there's no ACK to spoof. The client would receive both sets of packets and not know what to do. Initially, this seemed to make it nearly impossible to accurately replace content. At this point, inserting a RickRoll over other video streams seemed like a lost dream.

However, in order to get that video stream to play in the browser, there are a whole host of other things on the page that are transferred in the clear that can be replaced and made to help this effort. The most likely candidate is Java Script.

Replacing Java Script

Most Web pages today with any sort of active content (Web 2.0) include Java Script to drive the page and dynamically generate the content. Sites like YouTube drove the Web 2.0 movement but so did many blog frameworks and other user-generated content sites, most of which involved heavy use of Java Script.

The attack vector changed – instead of injecting whole videos, which was prohibitive, we replace some of the Java Script on the page, which is much smaller and easy to inject, and have it do all the hard work for us. In one test, a popular blog "failblog.org," which lists various pictures and videos of failures of common sense, was injected with a Java Script that sought out all links to flash video files (.flv files) and replaced the link with that of the RickRoll video. A page containing six videos was made to show the Rick Astley video for each. Only the embedded video was changed. Further tests on YouTube under ideal circumstances had every video, no matter which was viewed, play the RickRoll while the rest of the site was unchanged from the expected video content. At best, the malicious Java Script now links every YouTube video to the RickRoll. At worst, it replaces it with something less innocuous.

The ability to replace videos was deemed fun, but a larger issue started to emerge from this rather frivolous line of research. Java Script could be appended to the end of a file rather than to replace an existing script. All Airpwn had to do was watch the traffic stream for a </BODY> HTML tag and the attempt to beat the next packet, which is likely a FIN packet closing the connection. If Airpwn beats the FIN, the connection can remain open indefinitely now that they have hijacked the stream and the attacker can inject anything else he or she wants.

The implications of this were profound. Since the injected data is assumed by the browser to be legitimate traffic from the remote server, it is rendered by the browser. This has the advantage of giving full control over the contents of the page as well as interacting with the browser directly.

Recent times have revealed a whole host of vulnerabilities in browsers. There are too many to enumerate here, but most vulnerability announcements follow the same script; if you can get the victim to visit a Web page with the malicious code, the code will run and bad things will happen. The most common problem with this vector is getting the victim to visit the malicious site. This is no longer a problem since now we can bring the malicious site to them, injecting the malicious code into whatever site they are visiting. In addition, since the forged traffic is "expected," firewalls will do little to stop it.

This attack vector also opens up an interesting new vector of attack that can have long-term implications. In 2009, Robert Hansen (Rsnake) wrote a paper detailing some VPN vulnerabilities due to common internal address space usage and collisions of those addresses (www.sectheory.com/rfc1918-security-issues.htm). In his paper, he noted that one could attack the secure VPN network through the use of Java Script obtained from an insecure network.

An attacker who injects malicious code via an unsecure network has control over how long that script is cached; an attacker can force it to be cached permanently. The result of this is that the next time that page is visited, the cached version of the script runs, not the one from the remote server. If you did this for a commonly visited site like Twitter.com or Google.com, the script would conceivably run every day, on whatever network the client was connected to, even the secure VPN or even the "secure" local LAN at the office. Use of the client device on an insecure network, even once, is enough to open a vulnerability to the secure network.

This scenario has some pretty dire implications. Imagine setting up this attack at an airport or conference where the audience was full of employees of your target companies. Even if you managed to compromise just a few of them, when they go back to the office, they bring back the malicious code with them and it is inside the secure network doing your bidding. This could be everything from injecting a persistent iFrame into every page visited and sending an attacker the form contents, to acting as a stager, waiting for the right zero-day vulnerability to come out, then downloading and executing it automatically. Combine other attacks like Dan Kaminsky's Socket/Sucket DNS rebinding attack, which turns the victim browser into a proxy server for the attacker to access the internal network (http://events.ccc.de/congress/2007/Fahrplan/events/2393.en.html) and you have a formula for a very large amount of damage.

On an even more basic level, data injection need not be limited to Web browser content. As had long been postulated, with early versions of Airpwn but never implemented, DNS can also be injected. A client sends a DNS request for mybank.com and the attacker can redirect the client to their own phishing site. This can be a problem for an attacker since the range is limited and your target pool is small. You also have to know what site they will visit, but if you're in a situation where you want everything to resolve to a certain site (perhaps with extended browser exploits or other malware) or if you are at a conference or some event where you know attendees will likely be visiting a site for schedules or such, this could be very damaging. The other option would be not to spread malware, but disinformation. Imagine targeting a conference of bankers and rewriting all new sites DNS requests to show a fake news story about the economy melting down. The ensuing panic may be useful to an attacker, or at the very least, humorous to someone.

Injection attacks are limited only by the attacker's imagination. The potential damage can be either overt or subtle. The ability to inject and carry malware back to secure networks is probably the most troubling of these subtle attacks.

Normal methods of preventing these attacks usually come down to "use a VPN." The problem with that method is that many public hotspots use a captive portal, which you need to enter login information before the VPN tunnel can be established. This necessitates connecting to the access point and the captive portal over an unencrypted link. Even if the captive portal is SSL enabled, accessing the login page is usually done through DNS redirection of any page you view in the browser to the domain of the captive portal. This is enough of a gap to inject content since the initial connection can be hijacked and the redirect ignored.

Just as Karma was integrated with Metasploit in 2008, Airpwn is integrated into Metasploit as of late 2009. This means that Airpwn is now surrounded by a framework that will make it easy to build an inject content and also to utilize existing exploits to attack clients. The result of this will hopefully mean increased attention that using public hotspots or any unencrypted network could be a significant threat to secure networks long after the attacker has left.

SUMMARY

While all this may seem depressing, it is the inevitable cost for the convenience of wireless. While indispensible to modern life, wireless networks are far from the mature technology of wired networks. It took many years to convince people to even hook up a basic firewall to separate them from the Internet; it is not unreasonable to think that it will be a while before things improve for wireless security. As with most things, when systems were locked down at the access point, attackers just switched focus and found another weakness.

Attention to the dangers of using open networks is the first step in solving the problem. Usage of restrictive security policies preventing the use of insecure connections, while inconvenient, will help users protect them from themselves. As well, moving to more robust and intelligent client managers can also help alert users to the dangers and keep them from doing things that are blatantly dangerous. The last measure that is obvious but often not done is to turn off wireless and just use a wire when you can. It's a lot harder to screw up a wired connection than it is to screw up wireless.

In short, the best methods for avoiding attacks on wireless clients are

- Avoid using open access points at any time, particularly in public places.
- If at all possible, opt for a wired connection.
- If you are using a wired connection, disable your wireless adapter.
- Have firewalls in place and set to only allow services that you need.
- Avoid sending personal or financial information over wireless networks.
- Consider a "disposable" laptop for travel, one that gets wiped and reinstalled before connecting back to the company network.
- Always be wary that what you are connected to may not be what you think it is.

Endnote

1. www.wigle.net/gps/gps/main/ssidstats; [retrieved 01.12.09].

Bluetooth Attacks

INFORMATION IN THIS CHAPTER

- Bluetooth Technology
- Hacking Bluetooth
- Connecting
- Wholesale Sniffing
- Bluetooth Viruses

Bluetooth is one of those technologies that have become so common that it has become a part of our daily lives. It has become a solution to problems like driving and talking on a cell phone and introduced new and interesting marketing opportunities for attacks.

An interesting scam emerged that combined a Bluetooth attack with some creative usage of premium rate services.[A] Its effectiveness has diminished recently, but the effects are interesting.

Imagine a commuter is traveling home by train every night. As he or she walks through the train station, he or she is concerned about getting home in time for dinner, not about getting scammed via the phone in his or her pocket. An attacker is sitting near the waiting area, scanning the air for advertising Bluetooth devices. He or she spots the phone in our commuter's pocket and due to poor security choices is able to connect to it with a common or default PIN code. From there, the attacker instructs the phone to silently call out to a number, which stays connected until the commuter notices the phone call and dismisses it as accidental or the call is interrupted from a dead battery or service interruption. The attacker makes his or her money 1 month later when the commuter's cell phone bill arrives. The number called was a premium rate line owned by the attacker that charges a high cost per minute ($3.99 or some such number). The victim's phone called the number and incurred charges by the

---

[A]http://youtube.com/watch?v=cZRdnQ4g4NQ.

minute on his or her way home for the night. Over a 45-min train ride, that charge can grow to hundreds of dollars that goes into the attacker's pocket.

BLUETOOTH TECHNOLOGY

The need for Bluetooth was originally as a way to clean up our desks and decable our lives by being an alternative to the common RS232 cables that were needed to connect many devices to our computers. The ability to connect one device to several others simultaneously along with the ability to autoreconnect was a very exciting prospect for consumers tired of having to deal with many different cables for every device they owned. A single wireless method was needed to clean up the clutter.

Bluetooth is not actually one protocol, but it is actually a suite of protocols and functions wrapped up in a stack. Bluetooth operates in the 2.4 GHz ISM band along the same as 802.11b/g/n Wi-Fi; however, due to low transmit power and range, there is relatively little meaningful interference.

Bluetooth devices come in three classes. Each one is meant for specific applications and situations, but they are all the same in terms of compatibility. Most aftermarket Bluetooth adapters you can buy are Class 1 or 2. Class 3 is reserved for small devices like keyboards and a mouse that are sure to be near the computer they are connected to. Table 3.1 shows the relative power and range of each class.

Bluetooth has also undergone some revisions and multiple versions of the protocol exist. Bluetooth 1.2 and 2.0 + Enhanced Data Rate (EDR) are the most common in the marketplace, and you are unlikely to run into any of the 1.0, 1.0 B, or 1.1 devices except perhaps on some legacy and early devices.

Bluetooth 1.2 supports approximately 725 Kbps throughput, which is enough for things like mouse, keyboards, printers, headsets, and so on. Revisions to the standards culminating in Bluetooth 2.0 included EDR, which bumps the maximum throughput to 3 Mbps on compatible devices for larger data such as photos or music files.

Recently, the Bluetooth Special Interest Group (SIG) adopted the core specifications for Bluetooth 3.0. This specification works with the 802.11 radios in many products for a maximum throughput of 24 Mbps. Actual products based on the specification should arrive on the market in the beginning of 2010.

Table 3.1 **Bluetooth adapter classes**

Class	Transmit power	Effective range	Typically found
Class 1	100 mW	100 m	External USB adapters with external antenna
Class 2	2.5 mW	10 m	Internal laptop adapters
Class 3	1 mW	1 m	Cell phones, headsets, keyboards

EPIC FAIL

While Bluetooth-enabled phones are now commonplace throughout Europe, the adoption rate has been slower in North America. Early on, many North Americans were unaware of all the capabilities a Bluetooth-enabled phone offered since they have never actually had one with all the features. While not indicative of all models sold, many North American cellular providers selling Bluetooth phones were selling ones that had features disabled. Things like OBEX data transfer are disabled in the firmware and unavailable. The official reasoning was as a security precaution due to the issues described in this chapter, and some argue though that it is meant to preserve a revenue stream.

Most phones now have a camera built-in and the capability to use custom ring tones. These features are a great source of income for cellular providers. If you take a picture with your phone, you have to transfer it off if you want to upload it to the Web or share it with others. This requires an expensive data cable (sold by the provider), expensive removable media (if available on the phone, and sold by the provider), or uploading over the cell network via MMS or data, which costs airtime or additional charges.

The ability to pair a Bluetooth phone to a computer and download pictures or upload ring tones without additional charge or even wirelessly tether is quite appealing to the consumer. When so many computers come with Bluetooth capability, the user has everything he or she needs to transfer data without additional charge. This is why many cell phones are not capable of more than operating a Bluetooth headset or as a virtual dial-up modem.

In 2004, Verizon released the Motorola V710 phone. It was advertised as Bluetooth capable but controversy arose over the fact that Verizon failed to mention that most of the Bluetooth profiles were disabled via firmware, and the phone was not quite as advertised. A class action lawsuit was launched and eventually settled in September 2005.

The moral is that if you are buying a Bluetooth-enabled phone, make sure to check to see if all profiles are available if you want them.

Bluetooth devices are connected through a process called *pairing*. The pairing process usually involves one device searching for other devices in the area and then selecting the device to partner with based on its BD_ADDR (similar to a computer Media Access Control [MAC] address) or a logical name. Once pairing is completed, the devices are now considered "bonded."

Depending on the action being taken, a PIN for the device being connected will be required to complete the process. The PIN acts as a password in an encryption scheme between the two devices and is used to generate a link key. This key is used to secure communications between the devices and also to authenticate devices in the future. Should a device want to reconnect in the future, the devices check if each other's BD_ADDR is already bonded, and if it is, they assume that both have the link key still stored and will immediately begin communicating. This skips the PIN entry process and allows for automatic reconnection when in range. If one or both of the devices have forgotten the link key, the pairing process is restarted, and the PIN is required to be entered again.

Even before they are connected, devices communicate their capabilities to each other. Just as not every host on the Internet is a Web server or a Secure Shell (SSH) server, not every Bluetooth device talks on every service available. Through a protocol called Service Discovery Protocol (SDP), the devices announce to each other what services they support. A computer may support all possible profiles such as networking, file transfer, and others, whereas smaller devices only support a few or a single profile. A headset, for example, does not support networking and only has the audio profile available. This announcement of profiles allows devices to filter by capabilities, such as a phone looking only for headsets or a computer looking for a printer rather than being flooded with unnecessary devices.

Most often, the process is fairly painless to the end user. Some profiles can operate without encryption, say if you want to send over contact information from one phone to another without complicating matters with PIN numbers. Those that require encryption only require simple PIN numbers to complete the process. The addition of remembering bonded devices and automatic reconnect makes it even easier.

HACKING BLUETOOTH

With the widespread adoption and convenience of Bluetooth device comes the inevitable implementation problems that cause unexpected things to happen.

Most Bluetooth-based attacks are based on a simple and common flaw. Users often are very poor at reading documentation, at understanding risks and threats, and generally at changing defaults. Most attacks revolve around users not changing the default settings on their devices. That, coupled with poor user interface (UI) design, creates situations where users are unaware something bad is happening or about to happen. As well, documentation until recently did not fully explain the risks of certain actions.

Bluetooth Discovery

As with most attacks, the first thing to do is to find a target. Most often, Bluetooth attacks are against targets of opportunity (that is, not targeted). In the case of Bluetooth, its design assumes that devices interact with one another occasionally (that is the whole point of the technology). The need for these devices to find one another easily is a requirement of this. This allows legitimate users to find the device they are seeking, but also allows a nearby attacker to find those same devices and silently interrogate them to find out if they are suitable to attack.

The Bluetooth discovery process involves both parties in the pairing process. A device seeking to connect sends out a broadcast to the immediate area. Devices in the area, if they are set to be discoverable, then respond with the BD_ADDR of the device. From there, the querying device can use the SDP to discover all the profiles (that is, services) the device is offering (and vice versa). Simply put, a cell phone sends out a query, the headset responds saying it supports the hands-free profile, and the cell phone now knows to treat this remote device as a headset and not a dial-up adapter or some other profile.

Most cell phones, PDAs, or other Bluetooth devices have the capability to scan for discoverable devices. Even with this very simple interface, you can glean some interesting information. Many phones by default have the make and model of the phone listed as the name of the device when queried in a scan. This automatically makes life easier for attackers because they can immediately see what device it is and determine if there are any specific procedures needed to do bad things to it. Alternatively, when people do change the default settings, they very often change them to their name or some other very easy identifier like "Bob's phone" or "Alice's Blackberry." If an attacker is specifically looking to target Bob, then Bob has made his phone obvious among the others detectable in range and thus an easier-to-find target. By the same token, if the attacker knows Bob has a Nokia phone and sees only one device named Nokia 6330, then it's pretty obvious which one is Bob's.

The BD_ADDR, much like MAC addresses for network cards, has a known format to follow. They are 48-bit identifiers in a HEX format. The first 3 bytes are assigned to specific manufacturers. Like MAC addresses for network cards, the Institute of Electrical and Electronics Engineers (IEEE) assigns blocks of addresses to manufacturers to embed in their devices. The last 3 bytes are the unique address of the device. Since the assigned addresses are public, if the device name is not helpful, the BD_ADDR is not hidden and can be used to determine manufacturer. Anyone can query the OUI database and see what manufacturer made the device and adjust your attack accordingly. The IEEE has a query page up at http://standards.ieee.org/regauth/oui/index.shtml, and you can also download a copy of the whole database for integration into your own application or for local lookup.

> **NOTE**
>
> While manufacturers are assigned addresses in blocks that can contain millions of device addresses, that doesn't mean that all of them are used in the real world. Manufacturers may only use a few hundred thousand and then move on. The BNAP project aims to collect device addresses and see what actual manufacturer prefixes are in use. This greatly reduces the number of possibilities if we are searching address space since we can eliminate addresses not in use. The BNAP project's list of address prefixes and the manufacturers are available at http://802.15ninja.net/bnapbnap/.

Applications like BTscanner for Linux or Windows (www.pentest.co.uk/cgi-bin/viewcat.cgi?cat=downloads) or Bluescanner for Windows (https://labs.arubanetworks.com/) can automate the process of scanning for devices, interrogating and cataloging the results. This allows attackers to sit and observe the results much like Kismet (www.kismetwireless.net) and Netstumbler (www.netstumbler.com/downloads/) did with Wi-Fi networks.

There have been several projects over the years to quantify the number of Bluetooth devices in any given area. In early 2006, F-secure along with secure networks created the Bluebag, a mobile Bluetooth detecting rig including nine Bluetooth adapters built into a hard-sided wheeled suitcase. They wheeled the case through airports, shopping malls, and even a security conference. In 23 h of scanning at various locations, they discovered 1405 unique discoverable devices, many of which were advertising various profiles such as OBEX file transfer and headset capabilities. These devices may be susceptible to various attacks. The final report is interesting reading and is available at www.securenetwork.it/bluebag_brochure.pdf.

Another project to catalog Bluetooth devices has been going on since 2007 in the Netherlands. The Bluetooth tracking project (www.bluetoothtracking.org) has set up multiple sensors in various parts of the Netherlands and even Paris, France, that continually monitor for discoverable devices and record the information in a database. Since then, they have collected over 600,000 devices and generated some interesting usage statistics. They also have been able to calculate the speed at which a device is moving. One sensor detects a device, another a few miles away detects it again – from that you can extrapolate the speed at which someone was moving and what direction.

Some devices, as a security measure, do not respond to probes or only have limited window of time where the device is discoverable. These devices can cause a bit of a problem for the attacker; however, they are not impossible to detect with a little coercion and brute force.

Devices made nondiscoverable still can connect to other devices, except that the other device needs to know the BD_ADDR. If Alice wants to connect to Bob and Bob's phone is not discoverable, Alice needs to know Bob's BD_ADDR before she can connect. If she does not know it, Bob needs to make his device discoverable or tell Alice his device's BD_ADDR, and she can enter it (if possible) to specify what to connect to. When Bob's phone sees a directed request to connect to its address, it allows the connection to continue.

From an attacker's perspective, this makes it much harder to attack Bob's phone since it won't respond unless the attacker knows the BD_ADDR. It's unlikely Bob would share that with just anyone, but by using some basic information, we can potentially find out Bob's BD_ADDR.

Programs like Redfang (www.net-security.org/software.php?id=519) written by Ollie Whitehouse allow you to force connection requests through a particular address space. If we know Bob's phone make and model, we can either scan a similar model to determine its address range or search the IEEE database or BNAP database for that manufacturer known addresses. This will get us the likely first 3 bytes of the address. The remaining bytes will have to be brute forced sequentially until a response is received from Bob's phone. Since each query can take around 10 s to be thorough in waiting for a response, the process can take a while, from minutes to days depending on the type of device and the address space being searched.

CONNECTING

Once a device has been found, the device can begin connecting. In most cases, this involves an encrypted link be established before higher level communications begin. While there are variations, most often the device requesting the pairing is required to enter the secret PIN code from the other device. The theory is that only those who know the PIN of the remote device can connect. This may mean asking the owner of the other cell phone or just the fact that you own the device means you set its PIN and know it. Other variations include having the second device enter a code which has to be entered on the first device as well, thus requiring user interaction and avoiding default PINs. Once the PIN is entered, the devices handshake, and a link key is established, and now the devices are bonded and trust one another. This is an over simplification of the internals of the process, but the main issue here is not one of internals but of externals.

The most common attacks on Bluetooth devices are due to default PIN numbers. Many devices come preset from the factory with a PIN that the user is asked to change. Usually those instructions are buried in the manual and are ignored by the user. Other devices, such as headsets, have no interface in which to change the PIN, so their PINs stay static. The most common PINs are 0000, 1234, and 9999. A quick Google search for a specific device will turn up the manual, which will usually contain the default PIN, right next to instructions on how to change it. This situation leaves plenty of room for an attacker to play. The most famous of these attacks involving default PINs concerns Bluetooth headsets.

Carwhisperer

It's hard to escape them. Bluetooth headsets have become incredibly commonplace in daily society. The adoption rate has recently skyrocketed due to local laws that ban using a hand-held cell phone while driving. For others, they have been elevated to fashion accessories. Needless to say, as long as Bluetooth phones remain popular, these devices will continue to flourish.

Headsets are also related to the increasingly popular in-car Bluetooth hands-free devices. Occasionally integrated in vehicles by the manufacturer, and also available as after-market units, these systems take advantage of the car's built-in stereo system to allow the driver (or passenger, as it may be) to hear the audio from the phone while keeping both hands free (hopefully to drive). A small microphone is used to relay voice of the driver to the phone and to the other party – essentially, the same function as a headset, just not as fashionable, and not the easiest to wear on your head. For the remainder of the chapter, both in-car systems (manufacturer installed and after market) and wearable headsets will be discussed as the same thing since they are functionally identical.

Early Bluetooth security research noted that the usage of default PINs was going to be a problem since users, given the option, often not change a default setting unless forced. This led to carwhisperer being developed to exploit this.

In July 2005, carwhisperer was released by Martin Herfurt of the Trifinite group (http://trifinite.org/trifinite_stuff_carwhisperer.html). This software took advantage of the default PIN situation to provide a fun and sometimes frightening prank.

Carwhisperer is a series of scripts along with the actual application. The first cw_scanner probes for devices with the device classes common to headsets and hands-free devices. Once one has been found, the carwhisperer program initiates a connection to the device, and with the help of the cw_pin script, selects the appropriate default key based on the manufacturer prefix for the device's BD_ADDR. Once the connection is made, the attacker can record audio from the device's microphone or inject audio of his or her choice into the speaker. One can imagine the fun that can be had with the latter.

Once connected, the attacker can continue to listen and inject as long as they remain in range and the device connection is not broken. Carwhisperer will not work on devices already paired and communicating, as headsets only support one connection at a time. While this may seem a problem, really it is not. Many vehicles support the feature, but the person driving may not have his or her phone paired at the time. The car's system defaults back into discoverable mode, waiting for a connection. With headsets, there are specific places where headsets will be on but not paired. Favorite locations are movie theaters and airports since both locations require you to turn off phones, but people forget to turn off headsets.

WARNING

Be aware that airports, airlines, and flight crews are very strict about the rules regarding mobile devices being used at the wrong time. Some airlines and airports allow devices to be used during taxiing after landing, others do not. Still depending on the airport and the port of call, mobile devices may be forbidden from use until well after leaving the plane. Violating any of those rules can bring about fines and legal entanglements with the authorities.

The best location to do any guerilla Bluetooth auditing is as you wait to get on the flight and the passengers from the arriving plane disembark and turn on their devices – best to not tempt fate, safety, and fines by scanning while in flight.

While carwhisperer's usage in pranks seems obvious, a more devious use is in low-cost industrial espionage. Often in businesses, the policy for high-level meetings is to turn off phones, either for security reasons or to avoid distractions. Often people will turn off their phone but not their headset. Turning off the phone will break the pairing, and the headsets will return to a state seeking something to pair with. This allows an attacker to turn the headset into a listening device (bug) in the meeting and potentially divulging sensitive information.

The most obvious and easy solution is to not use Bluetooth headsets at all. However, this is not always an option and neither is changing the PIN on the headset since there is no interface to do so. The most effective way to avoid carwhisperer eavesdropping is to simply turn off the headset when not in use and to not wear it when not in a conversation (it's hard to record audio when it's in your pocket). Hands-free systems, in cars, can disable the feature when not in use or not paired.

Bluebug

The opening of this chapter described an attack involving Bluetooth being used to force other phones to make calls to expensive numbers owned by the attackers. Typically these are not one specific attack, but more a description of the end result, however achieved. Bluebug is the name given to the attack goals; however, the methods can vary. Originally named by the Trifinite group, it describes any attack that gains access to the Bluetooth device through an RFCOMM (another Bluetooth service) channel connection. Tools like Bluebugger (www.remote-exploit.org/codes_bluebugger.html) and Bluesnarfer (www.alighieri.org/project.html) automate the process that can be performed manually with some basic tools included with most Linux distributions.

These attacks are possible because of poor implementation by manufacturers or through undocumented debug or diagnostic services being left open. Some Bluetooth devices, most often mobile phones, expose RFCOMM ports (wireless serial ports) to the world and do not require authentication. An attacker can locate and scan a device for open RFCOMM channels and connect via a serial terminal. Depending on the phone's inherent vulnerabilities, the attacker can have nearly unrestricted access to the phone's contents and services, including reading SMS messages, reading phone book entries, issuing Hays modem (AT) commands, and allowing the attacker to perform such actions as setting up call forwarding or initiating calls. Due to the fact that the attacker is interacting directly with the phone, the UI often will not react and give any indication that the phone is doing anything.

To use the example at the beginning of the chapter of a commuter being attacked as he or she waits for the train, the attack can be accomplished with little more than built-in utilities and tools included with most operating systems. Obviously scripts and purpose-built tools can make this easier, but it does nothing to change the effect of the attack.

The attacker, using btscanner or some other application, would scan for discoverable Bluetooth devices. The BD_ADDR would show that the victim's phone is a Nokia. The name of the device is Nokia 6310i, which tells the attacker this is likely a vulnerable phone. Using sdptool (a utility included with the Linux Bluez protocol stack), the attacker can browse the advertised list of services and see if anything looks interesting. He or she discovers that RFCOMM channel 17 is open, and uses Bluebugger or Bluesnarf, or even the included rfcomm tool and serial terminal, to connect to the port and begin issuing commands. From there, the attacker can issue a command like "ATDT19001234567", which would tell the phone to dial a premium rate number that charges a large, per minute charge. If all goes well, the phone will connect and continue the call long after the victim has left the range of the attacker.

The worst case (or best case, depending on which side you are on) that will happen is that the target device will not be vulnerable and will require pairing before allowing connections to the RFCOMM port and may trigger a pairing request on the victim's phone, which will often be ignored or dismissed. Depending on the device though, if the victim is using a default PIN and his or her device does not require user interaction to pair, then all bets are off as the attacker can now connect anyway.

This attack causes the victim's service provider to act as the collection agent. The victim will have a hard time disputing the charge to the service provider since the call did originate from his or her phone and the idea of an outside attacker is often a hard sell. If the attacker is very creative, he or she will set up the phone line to have a very embarrassing name (some phone sex line or something of the like), so the victim is embarrassed enough at having to defend himself or herself that he or she didn't call

"Madame Whipsalot fun time party line" that he or she will just pay the charge to make the issue go away.

TIP

Some phones have open RFCOMM ports but are not advertised via SDP. These ports are leftovers from diagnostics used by the manufacturer and can be accessed by connecting to the channel directly. Depending on the phone, there are 60 channels available for simultaneous use. An attacker can just step through all ports to see if any respond. This is a time-consuming process, and it is more likely that he or she already knows which phones have these back channels open. If you are working to assess a Bluetooth device, it is worth stepping through all RFCOMM channels, just to be sure.

It is worth noting that a direct RFCOMM connection is not required to extract or abuse a target device. Other Bluetooth profiles can be abused to read information from devices. Profiles like OBEX push, typically used to push contacts from one device to another, can be exploited since the connection is actually two-way. Instead of pushing a contact, an OBEX GET request is made for a known filename. Files such as telecom/pb.vcf, which is the device's phone book, would be most useful to extract. This type of connection often does not require authentication. Some devices also can be caused to crash by pushing a vcard (a contact) with a very long file name, causing a buffer overflow and, possibly, crashing the phone.

At this point, it must be pointed out that many of these attacks have been mitigated through firmware upgrades and improved UI design. From about 2003 to 2007, these attacks were widespread as the technologies were new, and the user base was unaware of the risks. Many phones now are not discoverable, and most manufacturers have improved their implementations so as to not expose users to such risks.

That being said, there are always new devices coming on the market with Bluetooth capability. There may be an occasion where a manufacturer slips up and exposes a vulnerability. From the perspective of an IT or security manager, it would be a good idea to evaluate all Bluetooth devices and their peripherals for obvious vulnerabilities before allowing them in production environments.

From an auditing standpoint, tools like Bluetooth Stack Smasher (www.secuobs. com/news/05022006-bluetooth10.shtml) are invaluable as they have fuzzing capabilities to allow for discovery of previously unknown issues of security and stability.

WHOLESALE SNIFFING

In 2007, n.runs released Btcrack, a Windows program to crack Bluetooth link keys (www.nruns.com/_en/security_tools_btcrack.php). While this would seem to be a death sentence for Bluetooth, there is a problem. Sniffing the packets necessary to enable the crack to occur is a lot harder to catch than you would think.

As with any wireless technology, it all boils down to the fact that it is a radio, and radios operate in a shared medium, the electromagnetic spectrum. As such, its presence is impossible to hide. In order to communicate, it has to betray its presence. In the case of Bluetooth, this is the 2.4 GHz spectrum, the same as 802.11b/g/n networks; however, its operation is markedly different from Wi-Fi that creates some unique issues.

Unlike Wi-Fi, there is no monitor mode equivalent in Bluetooth since it's a suite of protocols rather than one cohesive package. When devices connect to one another, the master device of the piconet sets the hop scheme to all slave devices from its internal clock. Bluetooth operates using frequency hopping spread spectrum (FHSS) across 79 channels, each 1 MHz wide and hops at 1600 hops/s, which means that if we don't know the clock of the master, there is no easy way to follow the hopping sequence of the piconet. We won't be able to get all the packets. As well, it is not possible on a standard Bluetooth device to listen to all channels simultaneously, nor is it practical to have 79 Bluetooth radios each listening to a single channel.

NOTE

There are Bluetooth protocol analyzers on the market that do allow you to sniff all traffic. Typically these devices require the BD_ADDR of the master and slave devices and must follow the communication from the beginning, and even then, they can be tricky to get working right. These devices are not able to just walk up and listen in on random Bluetooth conversations like you can with Wi-Fi and monitor mode. These devices also cost thousands of dollars and are out of reach for most attackers and researchers alike.

An interesting thing about some of these analyzers, though, is that they often use the same hardware as consumer radios but different firmware. It was discovered in 2007[B] that some companies used the same radio chip in the hardware portion of their analyzer as consumer radios, the only difference being a single digit on the USB identifier and the firmware on board. It did not take long before people figured out that they could pirate the software portion of the analyzer, modify the firmware on cheap consumer Bluetooth dongles, and have access to an analyzer. Unfortunately for research, this was not of much help since, as noted above, you needed to know a lot more about the connection than a random attacker would in order to sniff all the traffic. However, this did point out that the hardware was capable, that the rest of the issues were software, and that it may be possible to build an open source protocol analyzer.

In order to sniff Bluetooth, we need to listen to the raw baseband signaling and all channels simultaneously to capture whatever traffic may be going by. This requires more than what a standard Bluetooth adapter can do. However, that does not mean that there is no hope.

The GNU Radio project was started to create a software-defined radio to allow people to freely hack the radio spectrum. Instead of specialized hardware to transmit and receive on a specific frequency, the GNU Radio went about designing a system with an field programmable gate array that could be reconfigured to any purpose.

[B]http://remote-exploit.org/research/busting_bluetooth_myth.pdf

The end result was the Universal Software Radio Peripheral (USRP), the companion hardware to the GNU Radio software, which allows researchers to reconfigure the device to behave like any radio device they need.

The USRP has some limitations for this end. The USRP2 (second-generation hardware) has a maximum receive bandwidth of 25 MHz, meaning that it can only read 25 Bluetooth channels at once, so you would need four USRP2s in order to scan all channels and capture everything.

Various projects like the GR-Bluetooth project (http://gr-bluetooth.sourceforge. net/) aim to turn the USRP into a baseband receiver capable of scanning all Bluetooth channels and capturing all traffic through various techniques. The project is ongoing and continually posts new code to its Web site.

In order to capture the necessary packets to crack the link key (the PIN is unnecessary to crack since it's used to derive the link key, and if you know the link key, reconnection is possible without victim interaction), there are a couple of options. The first involves impersonating the slave device in the piconet. The attacker determines the BD_ADDR of both devices. He or she then changes the BD_ADDR of his or her adapter to the BD_ADDR of one of the slave devices. The new slave tells the master that it has lost the link key, causing the master to request a new link key from the real slave. This is then captured and fed into Btcrack and out pops the link key with which can be used to decrypt the traffic captured from then on.

BLUETOOTH VIRUSES

Viruses are incredibly common today and are part of the background noise associated with the Internet and computing. While news of a new virus on a desktop computer rarely gets attention anymore, it is a surprise when it someone's phone that gets infected, particularly through Bluetooth.

Bluetooth has already been an infection vector for certain Bluetooth-borne viruses. The most well known of these viruses was the Cabir worm. In 2004, this worm surprised the world since it attacked cell phones instead of desktops. Targeting phones using the popular Symbian OS (SymbOS), the virus spread over Bluetooth connections automatically. Once a device was infected, no action was needed by the user to spread it; however, to be infected required a fairly large amount of user interaction. In this case, users would receive a request to transfer a file over Bluetooth from an infected device. They would then have to accept the transfer despite some devices warning against such actions. Once accepted, they would have to run the file, which would install the virus and begin propagation. The virus would then seek out other discoverable Bluetooth devices and continue the process.

Despite this obvious complexity, users still installed the virus and contributed to the infection. Often the infected device would repeatedly send transfer requests to the targeted device requiring the user to keep hitting "no." Often, the user would get frustrated with the incessant requests and would hit "yes" to find out what the fuss was about. Fortunately, various errors in the worm's programming limited its infection attempts to only one other device, keeping its propagation small and manageable.

This virus and others contributed to a small but growing industry for mobile device antivirus solutions. While not as common as its Internet-borne cousins, Bluetooth-based malware continues to be a potential issue that users have to understand and manage.

SUMMARY

Despite everything that you hear about Bluetooth security and its failings, it is not as bad as it seems. Many of the failings were initially due to poor implementations by manufacturers. As time has progressed, manufacturers have caught on about how to implement Bluetooth devices properly, avoiding these issues. However, with so many new devices coming on the market every day, it is very likely that issues will potentially creep back in, particularly from manufacturers new to making Bluetooth devices.

Care should be taken in any environment to ensure that users understand the risks associated with their Bluetooth devices. At the least, cursory testing should be done on any new device to make sure it does not have any glaring holes.

Ensuring that default PINs are not used and that Bluetooth functionality (at least discoverability) is properly configured can go a long way to limiting the impact of Bluetooth threats.

Radio Frequency Identification Attacks

Radio Frequency Identification (RFID) is a technology that everyone hears about but most people do not fully understand. Its name is thrown about in terms of efficiency and modernization of commerce, and at the same time, it is associated with Big Brother and privacy invasion. In terms of security, it is a technology that is only starting to come into maturity for security, but it is so often misunderstood and the risks so underestimated that security is often completely forgotten.

Most modern office buildings have done away with physical keys in favor of proximity card systems. We've all seen them, the cards on lanyards being waved in front of the readers at doors like some magical "open sesame" command for the modern age.

At the DEFCON 17 hacker conference, it is widely known that a fair number of attendees are federal agents, some under their own names and some undercover, but all there to learn. There was quite a stir among many of them this year: those who thought they may have had their proximity cards cloned. A project, undertaken by Adam Laurie (major malfunction) and Zac Franken to raise awareness about RFID vulnerabilities, gathered more data than expected.

Their project consisted of a 125-kHz RFID reader and a large antenna hooked to a laptop. The laptop also had a webcam attached to it. All of this was in plain sight on a table in the conference area. Anyone walking up to the table with a compatible RFID card on their person would have it scanned and their picture taken with the webcam. The idea was to then scrub the data and put it on the "wall of sheep," joining other attendees who used insecure protocols to login on the conference network.

It turned out that their project worked better than expected and may have captured the data from the door access cards of several federal agents. For conference organizers, this was stepping into a gray area (more than the usual for DEFCON) that they thought best not to encourage, and during the "Meet the Fed" panel, it was explained to the assembled feds on stage and the audience what had happened and the implications of this. At which point Adam Laurie destroyed the memory card with the data without having given it more than a cursory glance. Afterward, many were heard to say, "We never thought of that."

RFID BASICS

RFID encompasses many technologies under one acronym. Pretty much, any technology that is used to identify something via a radio signal fits under the RFID umbrella. As most people know it, RFID is a short-range wireless identification system typically used for building access, for barcode replacements, and in identity documents and payment cards. Chances are that you have one or more RFID-enabled devices on you at any given time. Most people, however, have little idea how these devices actually work, which like most technologies, leads to security risks and vulnerabilities.

RFID devices commonly have two parts: the transponder and the reader/interrogator. When the reader queries the tag, it replies with some identifier, most basically an identification number (there can be more complex responses which are discussed later), which is received by the reader and some action is performed with that identifier. This may seem overly general, but there is no end as to what can be done with a tag and reader. The tag can be placed just about anywhere, and the reader can pass the identifier to a back-end system, which can cause just about anything to happen, from an account being debited to opening a window or updating a map. The RFID identifier should just be treated as input into a larger system.

RFID Systems

Most common RFID systems fall into two categories: active and passive. This usually denotes the power system involved with the tag, as well as how it behaves when not being interrogated. RFID at its heart contains a very small processor that performs some action when powered; however, the method of power does limit some uses.

Active Tags

An active device typically has a battery or some other power system. The device is always "on" and typically has a longer range than passive tags. These devices are broadcasting their signal all the time and can be received by any appropriate reader in the vicinity. Major examples of this are the ubiquitous EZpass in New York State and other U.S. states in the Northeast and other toll collection devices around the world.

Usually, active tags are larger due to their requirements for internal power. There is no end of the sizes and shapes available, but they are typically much larger than passive tags. The advantage of an active tag is the increased range due to the extra power. This is one of the features that make them very attractive for toll collection since cars do not need to slow down or get unnecessarily close to the reader – the toll can be collected automatically, almost without slowing down from highway speeds. This reduces the manpower required for toll collection, as well as decreasing bottlenecks and traffic delays caused by manual toll collection. It also means that travelers do not have to worry about insufficient change since the tag is linked to an account that is billed later via credit card or invoice.

There is a third type of tag, the battery-assisted passive (BAP) tag. These tags require an external signal to "wake up" but once they are, they utilize a battery to vastly increase their range. These can be classified as active tags due to the use of a battery, but for some uses, it may be useful to consider them separately. For the purposes of this chapter, they will be considered in the active tag category.

Passive Tags

Passive devices are the other side of the technology. These devices are unpowered until brought into proximity of the reader. The reader powers up the tag through a process called induction. A radio signal of a specific frequency is emitted by the reader to the tag. The tag's antenna is tuned so that the readers' signal induces an electrical current in the antenna, supplying electricity to a small capacitor. When the capacitor is at full charge, it powers up the chip on the tag, which then performs its functions, sending out its reply through the same antenna to the reader. This process repeats as long as the reader is in range and is energizing the tag. Being that there is no requirement for an external power source like a battery, these tags can be extremely small. Some designs are just slightly larger than a grain of rice; others are thin enough to be made into flexible adhesive labels and attached to just about anything. Some packaging options can be embedded in clothing, namely the sewn in tag.[A] There are many packaging options available, with the tag itself only taking a very small amount of space in the package. Figure 4.1 shows a variety of tags and a small hobby reader.

Since passive tags are powered up by the reader, this requires that the reader be much closer to the tag than in an active system. The signal powering the tag from the reader decreases in power with distance (the attenuation is proportional to the inverse square of the distance from the tag), so to increase the read range, one needs to increase the power to the reader. However, the tag only has a limited range on its capability to send signals back to the reader, so if the reader is too far away, it may be able to power the tag, but not able to hear it. This distance is usually a few inches to a few feet depending on the specific technologies and readers used. This is not a problem for a door access card or a credit card system, but is not optimal for a car traveling at 50 mph.

[A]In many jurisdictions it is mandated that RFID tags in clothing be removable, so they are usually found as attachments to the cloth tags already in the clothing.

FIGURE 4.1

A Variety of RFID Tags and a Small Hobby Reader

Passive tags are by far the most common type of tag and are used in payment systems (credit cards), in authentication systems, such as door access cards, in counterfeit detection, such as tickets to the Beijing Olympics and passports, and on products of all shapes and sizes for inventory control.

RFID RISKS

RFID is often talked about as either a security panacea or a nightmare. It all depends on which side of the risk you stand.

As a panacea, its main benefits revolve around the capability for electronics to stand up to attacks by counterfeiters and other nefarious persons. They also provide much more potential of an audit trail and granular control in a variety of ways for things like inventory tracking and building access.

As a nightmare, they stand out because society in general has a tendency to trust technology. If someone can successfully counterfeit or clone an RFID device, it is assumed that it must be legitimate, and often there is no further check beyond the existence of a valid-looking tag.

The benefits of the technology are also its weakness. Since it has the power to transform any industry it is used in, it has the effect of people believing it to be infallible and as such, not preparing for its possible failure.

An issue that always seems to come up with regard to RFID systems and security is their expected longevity of use. A business investing in a system probably expects to use that system for 10 years or more and often it does. Should a vulnerability or unforeseen risk emerge in that time, they would most likely have to replace most, if not all of the system – something that costs a great deal and they are unlikely to do.

Another issue is complacency: the system has been operating for 10 years and those in charge of it may not be aware that there has been a new discovery that destroys all their previous conclusions about the system. It was a secure system when they installed it and still is as far as they know. Related to this is cost-breeding complacency, a business not wanting to opt for a higher-security system because they don't feel at risk (yet!) and can save money opting for the cheaper and less secure system.

In the security world, things can change, literally overnight. A once secure system can resemble Swiss cheese with one piece of information. With the speed of information dissemination around the Internet, potential attackers can learn of these issues in minutes. To correct an issue in the software world, this typically involves a patch or simply upgrading the software package to a new version or a reconfiguration. In a system where the vulnerabilities are built into hardware, this becomes an infinitely more complex problem to solve. RFID solutions are definitely in the latter category of a hardware problem.

In order to alleviate a vulnerability, all readers would have to be upgraded (via software if possible, but possibly the hardware too) and every card would have to be replaced. In an installation where there are hundreds of readers and thousands or millions of RFID-enabled card or devices, this is a huge task that will take significant amounts of time and money.

PHYSICAL ACCESS CONTROL

In today's world, it is a necessity that certain groups of people need to get into an area and other people need to be kept out of the same area. This has been an issue since the first humans built a wall to keep someone out of their space. If you were a prehistoric cave man and you wanted to keep creatures out of your cave (and hopefully from eating you), you built a wall of some sort so they could not gain access. This caused a problem, however. If critters could not get in, that likely meant that neither could you, and you may well become something's lunch without the protection of your cave. So shortly after the first wall was built, something resembling the first door was invented. This can be thought of as the first access control. The early human knew about the door and how to move it to gain access to the cave, but the critters did not. If you were a human, you could enter. This is physical access control; one person or group can pass through some physical barriers but another cannot. Over the years, this evolved into bigger walls, drawbridges, moats, guards, secret handshakes, and eventually to early locks and keys.

In more modern times, this grew into several different methods, all with advantages and drawbacks. Modern keys and locks can be fairly secure; however, granularity of access is difficult and may require users to carry multiple keys, which can be lost, requiring locks to be changed at great expense to prevent the lost key from working if it fell into the wrong hands. Mechanical locks also wear and break, requiring specialized maintenance, and in general, require a fair amount of overhead to keep the system secure in a large facility.

Guards are a common method of access control, even today. There are some major disadvantages to guards, though. They are expensive, they make mistakes, they can be bribed, fall asleep, or become an insider threat themselves. Guards also are not very keen on keeping audit logs of every employee's movements throughout a facility. That said, a human guard has some advantages and can greatly compliment a technological solution as will be discussed in a bit.

In 1960, magnetic stripe cards were designed by International Business Machines (IBM) and used as access controls by the Central Intelligence Agency (CIA). Mag stripe cards (like credit cards) have their identifier number encoded on a magnetic strip that is swiped through a reader connected to a computer backend. These systems are still in use today but suffer from problems of worn out cards and worn out readers. As well, since the read head for the mag stripe needs to be accessible and needs to be fairly precise, exterior access was problematic with dirt, rain, and other elements fouling the mechanisms. These systems, though, laid the groundwork for today's modern systems.

Finally it has evolved to what most modern office buildings are equipped with, RFID proximity cards (so named because the card and reader activate when in proximity).

EPIC FAIL

In 2006, the United States began issuing passports with RFID tags implanted in them as a security measure. The tags, when a proper key was provided, would respond with a digital copy of all the information on the photo page of the passport. The idea was to make it harder for counterfeiters since they now had to alter the physical passport, and the contents of the RFID tag to match the person presenting it.

When the plan was announced, privacy and security advocates were not pleased. Public comments were taken in and as a result, some changes were made to alleviate some fears, namely of the possibility of the passport being read at any time, not just at the border.

The solution consisted of a layer of metal foil in the cover of the passport acting as a shield against the tag sending or receiving any RF energy. This solution, on a technical level, works. When the passport is closed, communication with the tag is not possible. However, the solution is self-defeating. Due to the thickness and rigid nature of the foil and how it is embedded between the layers of the cover, the passport naturally wants to "pop" open about 1 inch, rather than stay closed like earlier generations. This gap is enough to allow the tag to be read, thus defeating the point of the foil in the cover. A passport in a purse or in a loose pocket will naturally want to open, potentially revealing data. This has lead to the common site at border control of wrapping an elastic band around the passport to keep it closed or using a special shielded case with a clasp for the passport.

This is a fairly common sight at most modern offices with employees simply waving cards in front of the readers that grant them access. You can usually see a small plastic box beside the door with one or more LEDs on it, though not always.

These systems can work via active or passive tags. Most of the older systems have a thicker active card, about the thickness of two credit cards that contains a coil antenna, the RFID circuitry, and one or more coin cell batteries. Other systems operate on passive tags, powered by the reader, and are often the same thickness as a credit card. There is no standard package for a proximity card system; however, the "card" format is most popular and is often integrated with the employee's ID badge as well.

Proximity Cards

The use of an RFID proximity system has many advantages over previous methods. Since it is a contactless system (wireless), the card and the reader can both be sealed against the elements. This allows for exterior readers to put up with rain and dirt and not be damaged. It also means that neither the card nor the reader can ever really "wear out" through repeated uses.

The most common and most basic use is through a simple numeric identifier. A card is presented to the reader and is queried. The card transmits its identification number, and the reader sends this to the back-end system. If the numeric identifier is allowed to enter that door, it sends back an approval message to the reader, which then opens the latch or door. If the presented numeric identifier is not allowed, nothing happens and the door stays shut. All of this takes place in a matter of moments and the user just has to wave their card in front of the reader to start the process.

Proximity-card-based systems offer many enhancements to security over earlier traditional systems like guards and keys. The biggest of these is the use of unique identifiers. In a traditional key-based system, everyone with a key to a door had the same key. Short of a secondary method to verification, there was no way to tell which key opened the door. This caused issues since everyone with a key could point at everyone else and say they were the one who opened the lock. There was no way to leave an audit trail of who opened what and when with keys. With a proximity card system, each employee's card has a unique identifier, which is tied to their name in the back-end system. This allows for discreet logging of a specific employee accesses and provides auditing, in case there are issues later.

The use of unique identifiers also allows for a high degree of granularity of control over what a proximity card can have access to. Traditional keys posed the problem, that to have really granular control over access, separate keys had to be issued for each lock and properly distributed. This was a real headache to manage and keep organized and led to people often having more access than necessary since it was easier that way. A proximity card system allows for very granular control and can allow you to specify down to individual doors, if you so desired.

If an employee is terminated, it is a very simple step to disable all his/her access at once rather than trying to change locks or reclaim keys that may or may not have been copied. If an employee loses a card, it is also easy to simply deactivate the old one and issue a new one and apply the same rights as the old one. A lost key in the mechanical lock system meant that locks would have to be repinned and new keys made and distributed.

Depending on the system, the cards can activate more than just doors. Machinery, cabinets, dispensers, vending machines, and even computer logins can be integrated with the system to allow a single card to perform many different functions.

Physical Access Control Failures

For a moment, it's worth looking at the concept of authentication since that is what this system is all about and where its failures lie. All the system is doing is taking the unique identifier and granting or denying access based on that identifier. This is where the problems creep in – it is just granting or denying based on the identifier. There is no mechanism to verify who is presenting the identifier.

There is an old saying within security referring to authentication: "Something you have, something you know, something you are."[B] This describes the different general methods a person has to authenticate to any security system, technological or not.

"Something you have" is simply a physical token that you must possess to gain access. The most common example is a ticket to a concert. You only need to have the ticket to get in: they don't care who is holding the ticket, just that you have it. The computer equivalent would be a hardware token like a smart card or a security certificate.

"Something you know" is simply a secret that only an authorized user would know. Most people know this as a password. From the login password to your computer to a kid's clubhouse or secret handshake to identify membership in a club, it does not matter who it is, as long as they know the password, handshake, dance, or whatever is being used to authenticate.

"Something you are" refers to biometrics. While this can be as complex as fingerprints, or retinal scan, it can be as simple as your friend, the club bouncer, recognizing your face and letting you in. Some unique element of you as a person is being used to authenticate into the system.

Generally within the security field, any system should integrate at least two of these things. "Something you have" and "something you know" could be a smart

---

NOTE

A special note should be made that while a traditional key is definitely "something you have", if more than one user has the same key, there is little point in using it as a security token. The logic of "Something you have" and of all parts of the authentication triangle is that every identifier is unique and thus can be traced back to a specific user, ensuring nonrepudiation. A common key among users means that now only one factor (a password, or biometric) is all that is left to authenticate.

[B]This is a security principal often taught as part of basic security classes. An example use is http://cs.cornell.edu/Courses/CS513/2005FA/NNLauthPeople.html

card and a password. "Something you are" and "something you know" could be a fingerprint scan along with a password. Any system should be built with at least two of these things (optimally, all three) required to authenticate to the system. This better prevents an attack on the system.

The corollary to the above saying is "Something I lost, something I forgot, something I was."[C] This emphasizes the failings of each of the individual authentication methods given above and highlights why they should be used in concert.

"Something I lost" is fairly obvious. If a user is issued a physical device to authenticate, he or she can lose it. Losing a key is the most obvious form of this that we've all dealt with. In the case of a proximity card system, a card can be lost or it could be stolen. As an attacker obtaining the card grants, you access simply by possessing the card, same as if you beat up someone in the concert parking lot and took their ticket.

"Something I forgot" should be very obvious to anyone who has ever worked technical support. Users will forget things. Passwords, handshakes – all can be forgotten and grind access to a halt. The subtext of this is that passwords can be divulged, sometimes innocently shared, written on a sticky note, or just guessed. Other times, they can be "discovered" through coercion, blackmail, extortion, or torture. Again, if you know the password and that is the only authentication method, then there is nothing stopping you once you know the password.

"Something I was" is something many people don't think about as much. This concept is a bit more of a gray area than the others. If your authentication method involves a biometric, it is hard to lose or forget your fingerprint or retina, but it also comes with new considerations. What about injury? If a fingerprint biometric is used, what if someone injures that finger or loses a whole hand? Are other fingers or biometrics enrolled? From an attacker's perspective, this one is harder since it requires physical access to the person they are trying to authenticate as. You might try to coerce the user to authenticate for you. You may try to forge or copy the biometric such as lifting a fingerprint and making a fake hand. It may also mean the more drastic step of amputation of a hand or finger to bypass a fingerprint system.

WARNING

An attacker removing a digit to authenticate as someone has happened. In Malaysia, in 2005, an accountant was car-jacked for his Mercedes S-class car. The car was outfitted with a fingerprint lockout device, enabling only the owner to start the car. After carrying around the owner in the trunk for a while to restart the car when needed, the thieves became frustrated and relieved the owner of part of his finger before dumping him on the side of the road (http://news.bbc.co.uk/2/hi/asia-pacific/4396831.stm). In any system, a determined attacker will find a way to bypass or defeat a security system, up to and apparently including amputation.

[C]As seen on a Shmoo group sticker at Shmoocon 2005 and mentioned among security circles ad nauseam.

The idea is that the difficulty and an attacker's level of commitment to the attack increase with each additional method. A password can be guessed remotely, but a password with a hardware token is much harder. Now an attacker has to physically interact with the victim, something most common attackers have no interest in. Using all three means that the attacker must have the password, the hardware token, and the biometric all present to authenticate, a fairly daunting proposition for all but the most dedicated attacker.

TIP

Some biometric systems use voice prints to identify a user. A person's voice can change over time, be affected by a cold, even by their mood, or by what they had for lunch. As such, to avoid high failure rates, these systems often have their tolerances tuned down, sometimes to the point where a recording of the user's voice will suffice to authenticate with the system. For obvious reasons, in high-security situations, this may be undesirable.

Any biometric system selected should take into consideration the failure rates and issues associated with a system and their suitability to the expected application.

Proximity cards' main failure is that they are usually a single factor form of authentication, namely the unique identification number. This means that, barring any other methods, an attacker just has to present the unique identifier to the reader to gain access. There is nothing that ties the unique identifier to the person it was issued to and whose access we are impersonating. There is also nothing in this type of system that ties the identifier to the proximity card it is expected to be on.

Cloning RFID

Since RFID proximity cards are wireless based, this gives an attacker an advantage over something like a physical key. Unlike a physical key that you would need to physically remove from the victim to copy,[D] the wireless nature of the proximity cards means that they can be read at a distance and do not need to be removed.

A large number of proximity card systems in use today, particularly older installations, only use a simple numeric identifier to authenticate. There is no encryption or any verification done on the card, other than to check the number received against a back-end system. This also means that there is no authentication by the card of the reader that is interrogating it. It will respond to any compatible reader and present its numeric identifier to it. This is very useful for an attacker since they can simply wave a reader past a pocket or purse of their victim and the card hands off the identifier. The attacker can then take this information and write his or her own copy of the card to use as they see fit.

In 2003, an engineer by the name of Johnathan Westhues took to tinkering with proximity cards and understanding how they work. Through his efforts, he was able

---

[D]While there are methods for duplicating keys from photos at a distance, this still requires the victim to remove the key from his or her pocket or purse to view it, limiting the time an attacker can obtain a picture. A wireless interface is available all the time for attack and just requires physical proximity.

to reverse engineer the protocol of the Motorola flexpass and understand what was being transmitted by the card and how it interacted with the reader. He continued his efforts and began to build a card simulator in order to interact with readers and understand how the cards worked. After a while and after some revisions, he built his own combination reader and card simulator, since both share many of the same parts. The device, called the proxmark, is not much larger than a credit card and only twice as thick and is dead simple to operate. It consists of two buttons, one for "reader mode" and one for "card mode." "Reader mode" turns the unit into a 125-kHz RFID tag reader that you can hold by any compatible tag and it will read its unique identifier and store it in memory. The "card mode" button simply replays the stored tag information. This means that he could walk up to someone and surreptitiously scan the contents of their proximity card, walk over to the secure door and replay their card, whereupon he would be granted access as if he had presented the real card.

Over the next several years, he has revised his device and greatly improved its capabilities. As of February 2009, the Proxmark 3 is the latest revision and supports most 125-kHz and 13.56-MHz tags (the two most common frequencies) and can read and emulate most any ID-only tag out there.

NOTE

All of the schematics, bills of materials, and designs for the Proxmark are available on Johnathan's Web site at http://cq.cx, if you want to build one yourself. If you are not inclined to build one yourself, completed devices are for sale at http://proxmark3.com/. A warning though that in most jurisdictions, cloning a card that you are not otherwise already authorized to have is probably illegal. You should only experiment on systems and cards you own or have express permission to experiment on.

The existence and availability of this device now means that an attacker can now clone many common proximity cards and simply replay them to gain access. There is still the issue of how to get to where we need to go in a building. If we are after secret research or something as valuable, its access is likely to be restricted to a few specific employees. This means that we cannot simply clone just anyone's card, we need to be sure to get one that will get us access to the area we need to in order to do our dastardly deed.

Role-based systems that allow granular access to certain areas seem like a decent way to limit employee access. They can only have access to where they need to be and can even limit access to certain times. However, an attacker can often use this to his/her advantage.

Optimally, these roles would be rigid. Only people with a real need to access a certain area would be granted access, and it would be strictly enforced. However, most businesses are a bit more political and pragmatic than that. Typically, those with legitimate access will have no limits on their access hours in case they work late, assistant staff may need to get access to do things on behalf of their bosses.

In a perfect world, Owner/CEO/Head honchos of the business have no real reason to go into some areas: their employees report things to them, and they don't need to actually go there themselves. People in these positions, however, tend to not like the idea of being locked out of anything or any limits put on them, and as such, their access is usually unrestricted and covers most, if not all, of areas and times.

As an attacker, we can use this by targeting the bosses and cloning their identifier. It can be as simple as sharing an elevator or in line for coffee. Sometimes, this does pose a problem though since the bosses of some organizations are not easy to access. There is one person, however, who will tend to have complete or nearly complete access to a facility: the janitorial staff. Just about any office has a cleaning staff who take out the trash, wash floors, and so on, typically at night; these people need to have their own access to enter areas to clean them. As an attacker, it's probably a lot easier to bribe or steal a janitor's identifier than the CEO's. One could walk up to a janitor as he is coming on duty and shaking his hand, and thank him for a good job, while your other hand silently clones the contents of the proximity card. Later that night, you can return and move about the building freely by simply replaying the janitor's card.

In either scenario, the audit logs (if any) will show the victim of the cloning as the one moving about the building. At worst (as an attacker), this will show the path you moved through the building. At best, it will direct the blame for your activities to the victim and away from you.

Minimizing the Risk

While it is possible to clone ID-only chips, that is by no means the complete end to using them for physical access control, but it did mean that it was not reliable as a stand-alone system. Many things can be improved in order to increase security of these systems.

One of the first methods to improve security is to actually audit the logs. Often, logs are ignored until something happens and then they are cursed for not being complete enough. Taking the time to evaluate if everything is being captured that you would need in a crisis will go a long way. In addition, automated auditing and alerts of the logs can be very useful in spotting inconsistencies and potential problems. If the enforcement of access controls is tight enough (scanning on both entrance and exit of a door), you can monitor for unusual activity such as Bob entering the building twice, or Bob's card being used at opposite ends of the building when there's no way he could have traveled that far. You can also look for situations where people are deviating from their normal schedules. A secretary should generally not be in the building at 3 A.M. Automated monitoring of the logs and alerts can help stop something before it starts.

In addition to auditing the logs, audit the enrolled cards. Periodically verify that only current employees are in the system and that access is as it should be. Employees who leave the organization should have their access immediately disabled but in case that does not always happen, periodic audits will help catch those situations. Also audit for test cards, often a new system will have one or several test cards issued with full access for the installers and administrators to test the system and make sure that readers work, doors open, and so on. Often, these identifications are not removed from the roles since the cards are destroyed. If the identifiers issued are sequential, it may be possible for an attacker to count backwards and generate cards from a known low access card to try and find some of the first issued cards.

As much as there are issues with security guards, they have one advantage in that they can think. If they are properly trained, they can add a formidable layer of security. Typically, a guard at an employee entrance will get to know the people coming in every morning. As an attacker, this is a real problem since we likely look nothing like the person whose card we spoofed. Some systems allow a human guard to see the logs of who is authenticating at the door and perhaps even an employee photo and may be able to spot the inconsistency of our access and detain us. This does require that guards not become complacent and be distracted watching TV or surfing the Internet, but it is a definite hurdle to overcome. Guards, however, are expensive and often their role is that of visible deterrent, rather than an active participant in the authentication solution. However, their presence can allow for more security than a completely technical solution alone.

Other ways to improve our fictitious scenario is to issue shielding sleeves for employees' proximity cards. These metallic sleeves block the signals from readers and prevent the readers from powering the tags, thus minimizing the risk from surreptitious cloning if employees actually use them. Actually explaining to employees the risks and implications of having their cards cloned can help get them on your side and prevent them from exposing the cards unnecessarily. Explain to staff that wearing cards on a clip or lanyard away from the office exposes them to unnecessary risk. Removing them and even placing them in a pocket will make them less of an obvious target. Other simple things like not leaving their cards in glove boxes where they can be easily stolen is common sense that occasionally needs to be reenforced by security and IT staff.

Even after all that, the risk still remains that cards can be cloned and that access is fairly easy to obtain. Often, these systems will be installed and in place for years and even decades. It's pretty certain to say that early designers and adopters never thought that their systems would ever still be in use when attackers gained the sophistication to attack them reliably. If the threat posed is great enough, upgrading or replacement of the system is about your only option.

RFID MEETS CRYPTO

One of the more obvious solutions for RFID as authentication is the introduction of cryptography into the process. The tag and the reader both have to prove a shared secret to one another before they will grant access. Cryptography has been used for decades in the computer world for authentication with great success. So why is it that it was not there in the first place?

RFID chips are small – very, very small. They vary depending on the make and model, but sometimes the actual silicon chip is smaller than the head of a pin. There is also very little processing power available on these chips. They are that way partially to make them easy to produce; it makes them cheap, and it also means their whole package will be small, which increases their applications.

With the diminutive size, it is a challenge to cram solid, known trusted algorithms. In the first few generations of tags, this just was not feasible. So manufacturers and designers did what they had to do in order to provide for the market demand for secure chips and broke a general security understanding, it's not a good idea to make your own crypto.

The most recent example of these types of system is the MIFARE Classic (http://nxp.com/#/pip/pip=[pfp=53422]|pp=[t=pfp,i=53422]). The MIFARE Classic system is a 13.56-MHz tag built by NXP Semiconductor. The low cost and availability at the time made it ideal for a great many applications, the significant of which were physical access control and fare systems for public transit. NXP boasts over 1 billion cards in circulation, accounting for about 70% of the market worldwide.[1] This, along with its usage as a fare system in London (Oyster card) and the Boston Subway (Charlie card) made it a very interesting target for research.

To secure communications, the MIFARE Classic uses a challenge-response authentication system. The challenge-response system is like a conversation you would have with a bouncer at a door to a club. You want to tell them the secret password number "5" to get in, but if you say it out loud, it could be overheard and anyone could get in. So you establish a mathematical system to prove that both of you knows the key without saying it out loud. This system must be hard for someone to reverse the process to get the number themselves. To keep it simple, the algorithm will be to take the secret number and the random number, multiply them together, and square the result.

You walk up to the bouncer and he picks a random number and says "My challenge number is 6." He has already calculated the expected result using the secret number 5. You calculate (in your head): $6 \times 5 = 30$, $30^2 = 900$. You reply to the challenge: "900" and issue your own challenge to see if he is actually the bouncer: "My challenge is 2." You already know that the answer is $2 \times 5 = 10$, $10^2 = 100$. If the bouncer replies with 100, you have both proven you know the secret and can proceed. If his response to you is wrong, you can walk away not trusting him or at the least, wondering about his math skills. If your response to him is wrong, he can throw you to the curb and move on to the next person in line. Fortunately for RFID, the results are less dramatic for failure. This is obviously an over simplified example as the encryption algorithms are (usually) much more complex, but the process is similar.

The MIFARE Classic system uses a challenge-response system to authenticate the reader to the card and the card to the reader without divulging the secret key over the air. The reader sends out a request for the unique identifier. The card replies with a 32-bit random number, the challenge. The reader takes this random number and computes it with the secret key both the reader and card share and the resulting 8-byte response is sent back to the card along with a 32-bit random number. The card finally replies with a 4-byte response to the reader and the unique identifier is now sent to the reader in an encrypted form. The card knows that the reader's response to the challenge should match with what it calculated and the same applies for the reader. This way both devices have authenticated to each other and the system can open the door.

The weakness was that the encryption system, called CRYPTO1, was built by NXP to fit on the chips in the tag and was proprietary. NXP made sure to keep the algorithm secret so that no one could understand exactly what was occurring. In the above example with our bouncer, it is necessary to keep our algorithm secret since someone who knows it and overhears the response can simply reverse the process and figure out what the key must be (square root of 900 = 30, 30/6 = 5) and then respond appropriately to any future challenge and gain access. This is a classic case of security by obscurity.

Originally launched in 1995,[1] the system quickly showed up everywhere and continues to be used today. It was not until 2007 that researchers were able to get anywhere in cracking the system. At the 24C3 conference in Berlin in December 2007,[2] Henryk Plotz and Karsten Nohl announced that they had successfully reverse engineered the CRYPTO1 encryption algorithm.

Their method was very labor intensive but ultimately successful. They took the integrated circuit (IC) from the tag, which measures approximately 1 millimeter by 1 millimeter, and very carefully scraped off the outer layers of the chip, exposing the transistors within. Then, with the help of a microscope, they photographed the chip in very high detail. This process was made even more complex by the fact that these chips are multilayered and they had to scrape off each layer individually, each only a few microns thick. After they photographed all the layers, they set about identifying the area responsible for encryption through some detective work. Once they had the proper section of the chip isolated, they used some image recognition software to identify the function of each transistor and how it interacted with others. After all this, they were able to reconstruct the internal logic at work within the chip that handled the encryption and reconstruct the algorithm. This breakthrough meant that they now understood the internals of the encryption and they could begin looking for potential vulnerabilities.

One of the interesting vulnerabilities they discovered was that the pseudorandom number generator (PRNG) in readers was stateless. Stateless meant that it did not keep track of its previous state between power cycles. These allowed them to power down and power up the reader and use the same pseudorandom number for multiple attempts, a big help in the encryption analysis world. It also showed that the random number generator on the tags was 16 bits, which means that it was a very small pool of random numbers and they were likely to repeat.

The reverse engineering of the algorithm was a huge step in being able to clone a card. Since compatible cards are easily available (anyone can buy them), it was just the matter of determining the key used on a legitimate card and transferring that to another card.

A few months after the presentation by Plotz and Nohl, in March 2008, a team from the Digital Security Group of the Radboud University Nijmegen in the Netherlands was able to recreate the algorithm themselves and began a cryptanalysis of the CRYPTO1 algorithm. It did not take them long to find several critical problems that led to the system being well broken. Their Youtube video caused a minor stir in the security world (http://www.youtube.com/watch?v=NW3RGbQTLhE)

One method involved attacking the reader. Using a modified proxmark3 board, they attempt 4096 authentications with the reader and record the results. Those results are fed into a 1-terabyte lookup file that contains all of the possible states of the algorithm, which also correspond to the keys. The attacker can simply generate this file in an afternoon and use it for any MIFARE Classic system they encounter. The results of the queries to the reader are looked up, which reveals the key.

Another and even simpler method of recovering the key involves no lookup table at all – it simply uses a weakness in the algorithm to give the attacker the state of part of the algorithm at the end of the transaction (even though it failed) and, with a little computing power, this is enough to determine the key for the system.

Once a key is recovered, the attacker can now use a mobile reader configured with the key to surreptitiously interact with a users' card, and since they authenticate properly, they can read all the information on the card. Once they have that information, they can write it onto another card and it will behave exactly as the cloned user. The Netherlands team demonstrated this with their Youtube video demonstration. In it, the attacker walks up with a laptop to the reader and using the proxmark3, collects a number of authentications and then returns to his accomplices who then use that data to recover the key. They then use the proxmark3 again (this time configured with the key) to wirelessly sniff the contents of a users' card. They then take that information and write it to a whole stack of cards. Each of those cards now authenticates as the cloned user.

NOTE

The cryptanalysis and internals of the attack could be a book in and of itself and is glossed over here. The original paper describing these attacks is available at http://www.sos.cs.ru.nl/applications/rfid/2008-esorics.pdf

---

Considering the number of cards deployed and that many secure facilities use this technology, some people were none too happy about this revelation. NXP was certainly not happy. The team planned to present their findings at the Esorics 2008 conference in Istanbul, only to be challenged in court by NXP who filed an injunction to prevent them from presenting their research. The courts eventually found in favor of the researchers and allowed the presentation to proceed.

EPIC FAIL

This scenario was later brought to the forefront again at the DEFCON 16 conference in 2008. Several students from Massachusetts Institute of Technology (MIT) were scheduled to deliver a talk about the Boston area subway system and several vulnerabilities they have found throughout, from physical to network, social engineering, and ticketing. The Massachusetts Bay Transit Authority (MBTA) took exception to the guerilla research done to their RFID payment card, named the Charlie card. The students had basically implemented the attack by the Netherlands group and had applied the attack to the MBTA readers. The MBTA filed a last-minute federal lawsuit against the students barring them from delivering the presentation. The Electronic Frontier Foundation (EFF) stepped in and was able to get the lawsuit dismissed, but only after the conference was over, which is what the MBTA seemed to want in the first place.

The students' slides were posted to the Internet by the MIT student newspaper at http://tech.mit.edu/V128/N30/subway/Defcon_Presentation.pdf. The court documents are available online through the EFF at http://www.eff.org/cases/mbta-v-anderson

SUMMARY

RFID is a tricky technology to secure. It is wireless, and as such, communications can be read at a distance and captured. Due to their small size, the options for encryption are limited and often rolled by the manufacturer themselves instead of peer-reviewed algorithms, which can lead to them being defeated with ease.

In the end, any system based on RFID should ask the following questions from the beginning: "What do we do if this fails?" "If a tag is cloned, what will the effect be?" "Is there another layer of protection that will minimize the impact if this occurs?" "What is the organizational structure around the use?" "Is it possible to audit its use and monitor for irregularities?" All are tough questions that must be answered.

RFID is a young technology, and developers and society are still adapting to what it all means. There is no simple solution and it is up to the integrator/user of the system to actively think about what the system's authentication scheme means in a grand sense. Simple authentication with a card may be good at limiting access to the front door of a building from the general public, but perhaps not the crown jewels. RFID-enabled credit cards may seem convenient, but one must weigh the convenience of this against the possibility of a drive by skimming of the credit card details. In that particular case (1 factor), physically swiping the card and requiring a signature is a better (2 factor) and more secure solution to keeping your credit card safe.

Endnotes

1. www.nxp.com/#/pip/pip=[pfp=41863]|pp=[t=pfp,i=41863]
2. http://events.ccc.de/congress/2007/Fahrplan/events/2378.en.html

Analog Wireless Devices

Many people are amazed by the amount of wireless devices available now. From mouse and keyboards to telephones, everything is being made wireless. It is common for people to interact with wireless devices dozens or hundreds of times a day. We hardly stop to think about it anymore (except when batteries fail). We often fail to assess the security of these everyday items because they are just that, everyday items.

While most modern digital wireless devices offer some level of security in their operation, people often fail to realize that many of the devices they use are still using analog signals to transmit data and offer little to no protection against outside eavesdropping or interference. Even as the first decade of the twenty-first century closes, we continue to use devices using the same technology as over 100 years ago and at just about the same level of security as back then too.

Most offices today have two things on every desk, a computer and a telephone. The computer usually requires two hands, and the phone at least one. Since humans have yet to adapt and grow a third limb, it is easier to adapt the phone. Out of convenience and comfort, most modern offices that spend any amount of time on the phone will often use headsets instead of awkwardly cradling the handset in their shoulder. Modern conveniences have done away with cords and gone wireless, reducing cable tangles and leaving both hands free to work the computer.

Many of these offices, however, never realize that their convenience is potentially bleeding out private, proprietary, or personal information to anyone who cares to listen. A penetration test was able to use these headsets to glean enough information to accurately impersonate a remote employee "visiting" the office. He or she is able to gain full access to the building, bring in accomplices under the guise of vendors, and even be assigned a desk with a handy network connection. He or she spent the next several days scanning the network for vulnerabilities all while consuming the free office coffee, all because of some information gleaned from conversations over wireless headsets.

Often overlooked, analog wireless devices can be a very large gap in the security of any office and must be taken into consideration as part of any security effort.

ANALOG DEVICES

A well-timed article in January 2008 on the security Web site, DarkReading.com, told a story that many penetration testers are all too familiar with, but it is just a sample of what happens every day and the problem of "we didn't think about that" security.

The article (http://darkreading.com/shared/printableArticle.jhtml?articleID= 208803553) by Steve Stasiukonis explains a scenario he experienced on a penetration test. The client asked his team to test all the radio emissions from the target building, including Wi-Fi, Bluetooth, and anything else the author could find. This was of interest to them as they were keen to try new things and were very interested in the cordless headsets in use by many employees.

They used a handheld radio scanner to scan through common frequencies for these devices and identify headsets in the target office. Once they knew the frequencies, they were able to eavesdrop on phone conversations and glean a large amount of information in a short period of time. Information they gleaned were things like participants in conference calls, employee names, locations, schedules, and other office gossip. It was not just phone calls either. Many headsets continued to transmit after calls were hung up, broadcasting office conversations as well. This information allowed him to impersonate an employee from a remote office visiting the target office. The plan worked frighteningly well. He was able to convince security that he is the employee with some dropped names of employees and a fake business card. He was assigned a desk and given a building access card right away with almost no questions asked. His new desk gave him network access and the freedom to sniff traffic and scan the network all day long.

After a couple of days, he was able to book a conference room and invite a "vendor" (an accomplice actually) in for a meeting. This allowed them more network access and the ability to bring in further attackers past perimeter security.

All of this was accomplished with just a little insider knowledge gleaned by listening to insecure transmissions. The amount of information they learned would have allowed them insider knowledge on trade secrets, stock plays, and other very private information.

The article caused a fair bit of interest because it was something that should be plainly obvious, but something very often overlooked.

DIGITAL VERSUS ANALOG

Most people are used to analog communications without knowing it. We've used it for the past 100 years quite successfully. Analog transmission is where a piece of data, be it voice or anything else, is sent with a continuous signal that varies in amplitude or frequency, much like wave rolling onto the shore is a continuous action. Small changes in the signal are meaningful to the data being transmitted.

Digital on the other hand is noncontinuous and uses discreet signaling to send data. Information is broken down into 1's and 0's and sent as discreet states. It's as if the wave rolling onto the beach was out at sea, then the next moment it was on the beach, then right back out to sea.

Analog has the advantage of more accurate representation of complex data such as sounds since it's a smooth transition from high notes to low notes and accurately represents the changes in air pressure associated with sound. The problem with analog is that noise in the medium can interfere with the subtleties of the transmission and the received message is garbled (in the case of sound this is "static"). Digital transmissions break down data into a digital representation of the original encoded as 1's and 0's. It is not as accurate but is much more tolerant of noise in the transmission medium. This is why Morse code is so effective. It has two distinct and easy to recognize states: dots and dashes, effectively 0's and 1's.

Most often, analog is used in short-range single channel type communications such as microphones and headsets since the circuitry is easier to make and the short range minimizes potential interference issues. The circuitry is also often cheaper than digital as well. Digital allows for more data to be sent in a smaller amount of the medium (that is, smaller bandwidth) since it only needs two distinct states to represent the data. This allows for more devices to operate in the same medium as opposed to analog.

Analog Security

Analog radio is not that easy to secure. Unless you process the signal beforehand (encoding it or changing it to another form), it's pretty easy for anyone to intercept the signal. Most people have heard stories about bored people using amateur radio gear to transmit their voice onto fast-food drive-through speakers. They overpower the employees' signal and replace it with their own. Usually these pranks end up resulting in childish and crude humor laced with expletives and some very confused employees and angry customers.

The most common way of securing data over analog is through methods like frequency hopping where the sender rapidly changes what frequency the message is being transmitted on and the receiver knows the order of these changes and can keep up. An outside listener who doesn't know the order only hears tiny pieces of the message and cannot easily reconstruct the whole message. The problem is how to ensure that both the sender and the receiver know the hop order and can stay in sync while preventing an attacker being able to determine the order and following along.

The other method of securing analog communications is to preprocess the signal, adding audio interference or other techniques before broadcast. There are few systems like this in current operation. A common example is early premium cable channels in the 1980s which inverted and injected content into the signal which paying subscribers needed a "descrambler" to view. Most modern systems choose to simply convert the original signal to digital and encode/encrypt that before transmission.

Digital Security

Digital security allows for a much greater number of options for security. The message can be encoded or encrypted before transmission much easier than analog systems since all that changes is the order of 1's and 0's. This can also be layered with analog methods like frequency hopping to further make eavesdropping difficult. Again, the difficulty is in making sure that both the sender and the receiver know how to encode and decode the information being sent.

CORDLESS AND WIRELESS

Cordless and wireless are terms used every day by most people, yet most don't stop to think about their meaning. Cordless is most often used in reference to some home telephones. In this case, cordless refers to the lack of a cord or wire between the base station and the handset. It has been replaced in this case, with a radio frequency (RF) transmitter. Cordless does not always have to denote a radio transmission however. Many people forget that there are other mediums that data can be transmitted over. Take the ubiquitous television remote control. It is a wireless device that trades cables for infrared transmission (nonvisible light energy) of data as opposed to RF signals. Other data transmission systems use lasers for line of sight data transmission.

It is worth pointing out that there is no specific designation as to whether or not a device that is labeled cordless or wireless is using analog or digital signals. Usually the manufacturer's datasheets, manuals, or Web sites will give some indication as to the transmission method in use.

EXPLOITING ANALOG WIRELESS

Analog wireless devices are usually fairly simple things that do one thing. Often, they are used where simplicity and quality are necessary. The complexity of security of these devices is often not considered worth it in many contexts; however, problems can arise when the technology is used in places it probably shouldn't.

Audio Vulnerabilities

Major threats from analog wireless signals come from voice transmissions to and from headsets. These headsets are often used for phone calls as a replacement for the handset but do not always need to be. Usually situations of performances or presentations can also necessitate the use of wireless microphones for amplification purposes. Also these may be used where quality is not necessary but ease of use is. Rarely is any thought given to the security of things said over these devices.

If you take an average day at the office and the phone calls made throughout, an amazing amount of information is passed via voice. Conversations between employees about what seem to be innocuous or minor things about the business could be of great interest to an outsider. A secretary receiving instructions from his or her boss to make some changes to a financial report could be terribly interesting to someone looking for an inside edge on the stock market.

In order to do business, information has to flow. Given that there is this huge flow of information in an office environment, it makes sense that it would be a target for attack. On a smaller level, this is why there are offices in the first place, for privacy of that information flow. In more extreme cases, this is why there are locks on the doors and guards at the front desk, to prevent an attacker from entering the area and eavesdropping on that flow.

Unfortunately the enemy of "security" is very often "convenience." Given the amount of time some people spend on the telephone, it's not unreasonable that they would want a more comfortable way to use it. Enter the headset market.

Headsets (also called hands free and other names) have been around a fairly long time. Photos of early telephone operators show rows upon rows of employees wearing headphones and microphones. Given they spent a long time on the phone as that was their job, holding a handset up to their head was impractical. The telephone operators eventually went away, but the headsets were adopted by call centers, secretaries, and anyone involved with heavy usage of telephone. As time moved on and radio hardware became cheaper, cordless headsets were introduced to remove the clutter of cables from desks. Without realizing it, they had now made it easy for any interested party to listen in on conversations, well outside the range of locks, guards, and/ or other security measures. The use of headsets, while it made the employees' lives easier, it often came at the cost of security.

Just about anyone who has ever dealt with them knows stories of baby monitors picking up conversations of neighbors who have similar units. Usually this is accompanied by idle curiosity or some voyeuristic intention to find out more about the people being heard. This is the basis of the vulnerability; an unencrypted radio does not care about who receives the signal, and anyone with a compatible receiver can receive these transmissions and listen in. In the case of baby monitors, just having a unit that operated on the same frequencies and being within range is enough to listen in on private conversations. The same is true of many fast-food restaurant drive-through ordering systems. The employee speaks over an analog headset to the speaker at the curbside, and a microphone replays the order back to from the customer. The speaker at the curbside does not care about who is transmitting on the same frequency, only which signal is the strongest. In the case of pranksters, it's not hard to obtain a radio that is powerful enough, and with some modifications, capable of broadcasting to drive-through speakers. Sometimes creative, sometimes not, these pranks can be entertaining to watch when documented.[A]

SCANNER SELECTION

There are a huge amount of scanners on the market. Some are aimed at very simple users, while others are aimed at amateur radio enthusiasts and others for professional use. Some of the simplest scanners are aimed at sports fans, namely NASCAR race fans, that use them to listen to the radio transmissions between their favorite drivers and pit crews. Others are aimed at those with voyeuristic intentions and are typically sold as "police scanner" in that they focus on frequencies in use by law enforcement. One does not even need a scanner in some cases as with the advent of the Internet and streaming audio; scanner fans have put their scanners online for anyone to listen to online for free.[B] There is even an application for the iPhone for listening to scanner feeds from many cities around North America.[C] While these are limited to police bands, they show the kind of information that can be heard right out of the air.

WARNING

When looking to purchase a scanner, be aware that local laws may ban their possession or use by those without special permission or a license (amateur radio). Other jurisdictions have bans on some frequencies and others on convicted felons possessing them. Some rules also cover handheld scanners as opposed to vehicle-mounted units.

These rules do not necessarily work to stop the criminal element that may use them against you since they won't care about the law anyway. However, as a law-abiding citizen wanting to test your own equipment, it is highly recommended to familiarize yourself with all the relevant local laws before purchasing or using a scanner or else you might find yourself on the business end of the law.

The most important thing to consider in selecting a scanner is the frequency range. A cheaper scanner for the NASCAR crowds may not be capable of the frequencies needed to listen in on your target device (but they may be, depending on the target), but they will probably not have the range necessary. Some research will be needed ahead of time to determine what frequency ranges you require. Generally speaking, the wider the range of frequencies supported, the higher the cost.

[A]http://phonelosers.org/tv/drivethru/
[B]http://police-scanner.info/index.htm
[C]http://scanner911.appspot.com/

Another consideration in selecting a scanner is the form. A large tabletop unit may have all the features you require, but it is not that easy to transport to the client site or use discreetly. Handheld scanners also deserve consideration for things like audio output capabilities to record signals, computer control capabilities, trunked radio tracking, and even battery life and external power options.

Additional features can be useful to consider as well. The size of the display, video capabilities (discussed later), and memory functions are all important to consider. There is no single unit that does everything you could ever want (there may be, but cost may be a factor), but most times you won't need all possible features, just the ones applicable to the job at hand.

EPIC FAIL

Early analog cell phones were fairly easy to listen in on with commercially available radio scanning equipment. Their transmissions were unencrypted and could be easily intercepted by anyone with a suitable receiver and listened to by anyone in range who was curious.
Of course, things said on a phone were often more interesting than things said around baby monitors leading to users asking questions about security. Regulators of different countries reacted quite differently to solve the problem of eavesdropping.

In the United States, the mobile phone industry, rather than improving the technology and adding security (which would add cost to an industry in its infancy), instead decided to press the FCC to institute a ban on any scanner capable of monitoring analog cell phone frequencies or any device that could be modified to receive those frequencies.[D] This did not actually prevent calls from being able to be eavesdropped, but just banned the tools to do so. After a certain date, all scanners sold in the United States were required to have those frequencies blocked out. Of course, the criminal element had no problems continuing to obtain unlocked scanners; the law simply prevented Joe Average from doing so. Effectively they were trying to jam the genie back in the bottle. This ban and lack of discussion in subsequent years about the realities led to many people not realizing that their cell phone conversations were vulnerable.

Canada, by comparison, took a different approach. The Canadian Radio-television and Telecommunications Commission (CRTC) says that listening is permitted; however, disclosure to others of what you hear, or usage for personal gain, is punishable by law.[E] This allows Canada to sell unblocked scanners. Pressure by the telecom industry did cause many to be sold with the cell phone frequencies blocked, but some were not.

Neither country's action prevented someone with nefarious intentions from illegally acquiring a capable scanner, listening in on conversations, and doing whatever with the acquired information.

While the move to digital cellular phones added support for encryption technologies, it is still up to the providers to enable them and use them. In some circumstances, they may choose not to do so, as the additional overhead can mean less users per tower and increased costs against their bottom line. For this reason, it is worth investigating other security options such as encrypted phones since the continuing ban on equipment to monitor cellular phone frequencies is unlikely to deter against a determined attacker.

[D]http://fcc.gov/Bureaus/Engineering_Technology/Public_Notices/1997/da971440.txt
[E]http://canlii.org/en/ca/laws/stat/rsc-1985-c-r-2/latest/rsc-1985-c-r-2.html#9.-1

Headsets

Many cordless headsets in offices today are digital and often not as vulnerable as their analog predecessors. That being said, there are still a great many devices in use due to the adage "if it ain't broke, don't fix it." These legacy devices can be used for years after others have been replaced.

NOTE

A cordless telephone can be thought of as a headset as well, albeit much more bulky, and much of what is discussed in regards to headsets can apply to them as well.

These devices (and the most likely to be vulnerable) will typically operate on the 46 to 49 MHz range or the 900 to 948 MHz range. Others operate on 1.2 GHz and higher frequencies but are often digital in nature and not subject to the same vulnerabilities.

Eavesdropping on these devices is as simple as obtaining a wireless scanner capable of receiving these frequencies and tuning in. Many headsets on those frequencies use simple frequency modulation, easily handled by most scanners. From there, you can simply listen to what is being spoken at that time.

Depending on the system, one frequency is used for base station to headset audio, and another is used for headset to base station audio. This means that you can only receive one half of the conversation. If you are only picking up one side of a conversation, try to identify the other frequency in use (usually only a few megahertz away).

Being able to eavesdrop on telephone conversations from the parking lot is a rather tantalizing prospect for an attacker. Taking a suitable scanner into the business centers of any major city will yield some sort of interesting conversation – conversations about upcoming financial filings or business deals, to tech support calls to reset passwords on user accounts. There is no limit to the kind of information that may be gleaned.

WIRELESS MICROPHONES

While related to wireless headsets for telephones, wireless microphones tend to be different enough to warrant separate discussion.

If you've ever been to a conference or presentation, you have probably seen these wireless microphones. Usually pinned or clipped to a lapel, there is a cord snaking to a transmitter in the pocket or clipped to a belt that sends the speaker's voice to the camera or to the room speakers so everyone can hear clearly. Hardly a presentation made to a reasonably sized room occurs without the use of a microphone like this. As well, stage productions will often use similar (though higher end) equipment to amplify the voices of the performers on stage so the audience can hear clearly.

These devices are a rich target for eavesdropping and disruption as they are most often unencrypted, and while the room may be closed to the attacker, the radio signals do not respect those same boundaries. Imagine a presentation by the CEO of a large company to the assembled stockholders where he has to inform them that their latest widget does not only work as promised but also has the side effect of turning customer's hair blue. The CEO knows that he must disclose this to the stockholders but is trying to keep it out of the press. He begins his presentation by clipping on a wireless microphone so everyone can hear him clearly, now including the nosy reporter across the street with a radio scanner. Usage of a scanner during closed door presentations can yield very interesting results during penetration tests and have proven to be very effective more often than not.

These microphones typically require higher quality than telephone headset units and operate on different frequencies and use much wider channels (200 KHz usually) so that voices are represented clearly and accurately. Most modern systems operate in the Ultra High Frequency (UHF) band, but due to different licensing and allocation for those frequencies, the actual range varies depending on what country the device is configured for. Typically though, you can find them between 470 and 854 MHz (698 to 854 MHz in the United States). If you know the manufacturer of the equipment, it is usually fairly easy to search online for the frequencies that the equipment uses.

EPIC FAIL

The future that is often predicted by science fiction is one where our lives are made easier by technology. Among those inventions often sought by business users are voice recognition and voice dictation.

While voice dictation products have gotten to the point of being fairly accurate, the quality of the output is directly related to the quality of the input. The common problem shared by all voice dictation software is how to get the user's voice clearly into the program so that it can work its magic. A poor quality microphone introduces too much noise and as such, the number of errors increases. Often this is not a huge issue as for most people they can go back and edit later. Other users do not have the time and require accurate dictation the first time.

A common solution for these users is to avoid the consumer-level microphones and step up to high-end stage equipment. This usually has sufficient quality to reduce dictation errors.

The epic fail is when those users are in fields where they are dictating private, confidential, or secret information. Every word they speak to their computer can be received by an outside party (who could even then pipe it to their own voice dictation software). Sometimes, these users will only disable the dictation software when not in use and continue to wear the headset and leave it on throughout the day. The implications of this are staggering as the contents of every phone call and conversation around the office are now being broadcast outside the walls of the building. Once this situation is noticed, the cost of the microphone is now wasted as there is no real way to secure those communications after that point and continue using the headset.

When using a wireless microphone, you must consider the content of the audio being sent and what its implications are. Treat the microphone as a stranger in the room and ask yourself, if you wouldn't talk about it in front of a stranger, why do you have a microphone broadcasting it to anyone in range?

VIDEO DEVICES

Video surveillance is ubiquitous today. Most major cities have hundreds of cameras around watching everything from traffic flows to back alley trash cans. Whatever the reason for installing the camera, most cameras have the purpose of allowing someone to monitor an area remotely. Sometimes, these cameras need to be installed temporarily or in places where cabling is not easy or practical to install. In these cases, wireless video cameras are popping up to fill the void. Many homeowners use these wireless cameras to monitor their yards and surrounding property, as well as temporary applications like construction sites. A hobby has even popped up to locate and catalog the locations of these cameras; much like wardriving, warspying emerged to attempt to better understand this new technology and how it was being used.

These wireless video units typically consist of a remote camera, either battery powered or AC powered, and a base station to receive the signals. Basic units are usually just a transmitter and receiver pair, whereas higher end units can have multiple cameras on a single receiver.

Video, due to its higher bandwidth requirements (than that of plain audio), typically needs higher frequencies to operate. While some early units operate in the 900 MHz band, these can suffer interference from cordless phones, wireless headsets, and many other devices and are often the cheapest available. More common are 2.4 GHz units, though these suffer from interference from Wi-Fi networks and other devices in the increasingly crowded unlicensed 2.4 GHz range. Often, they take up about half of the available Wi-Fi channels and cause more problems for Wi-Fi than Wi-Fi does for video. Newer units operate at 5.8 GHz, which is also populated with interfering devices like newer cordless phones, but more available spectrum means the interference is not nearly as much as it is in 2.4 GHz.

In order to pick up such broadcasts, once again a compatible receiver is needed. This can be a similar base station as the one in use by the target hooked to a video monitor. Some scanners have a video display and are capable of viewing video from these devices (handy on a blind assessment). A decent scanner will be able to scan through the common channels used by these analog video systems and display them. Others will automatically detect the video signal during a sweep and display it for you. Between radio direction finding and simple deduction based on the image from the camera, one can often determine where the camera is and what its coverage field is.

For attackers, these cameras are incredibly tempting targets as part of any information gathering operation. It allows them to literally spy on things going on in its field of view (guard movements, occupancy, even door lock combinations as they are entered!). Much of the common equipment on the market just broadcasts its signal, and no authentication is required to view it.

Much of this equipment is fairly cheap and marketed as "spy gear" in stores and online. While there are some legitimate uses ("nanny cams," employee monitoring, and so on), the low cost and the ease of use means that the capability to perform surveillance by amateurs is a very real and a danger to almost anyone. As a result, performing periodic sweeps for these types of devices with a radio scanner can ensure that no one has quietly installed one in your home or business. Professionally this is known as Technical Surveillance Counter Measures (TSCM), which is basically your classic spy bug sweep. Professional TSCM sweeps are very thorough and complete and depending on the type of organization, very necessary. Performing one on your own is cheaper though less effective, but good for that sound piece of mind at the very least, or vastly undervalued if you do discover something at the most.

NOTE

The reverse situation is true as well when attackers use audio or video equipment that is unencrypted that can work against their efforts and make them easier to detect and catch.

In August 2009, a news report from Oshawa Ontario, Canada, detailed how a 50-year-old man who was spying on his stepdaughter with wireless "spy cams" was caught when a contractor next door who was using similar equipment (for a backup camera on a large truck) accidently intercepted the video signal and determined the source and called the authorities.[F]

Had more professional, encrypted, or digital equipment been used, it is unlikely he would have been caught accidentally.

DEFENSE

The simplest defense against analog wireless vulnerabilities is simply not using analog wireless devices. With a little searching, most manufacturers' products can be easily determined to be analog or digital. Most often the devices with the greatest potential for damage can be the smallest. An employee suffering neck pains from holding up a phone handset all day with his or her shoulder to use both hands to type will often take it upon himself or herself to go to the local office supply store and pick up a cordless headset to make his or her life easier, usually without the knowledge or approval of anyone else in the office. Going from one end of the office to the other and reviewing every wireless device found, while time consuming, can ferret out these devices. At the same time, educate employees about the risks they may introduce and work with staff to find more secure alternatives.

Most digital wireless headsets sold today are Bluetooth based (which is encrypted) or even newer units are capable of connecting to and interacting with Digital Enhanced Cordless Telecommunications (DECT) base stations directly. These devices also offer much higher quality audio than older analog units and are a good option for comfort and ergonomics.

Wireless microphones are a bit of a trickier issue to secure. Much of the problem is that in most people's minds, it's not necessary to secure them as the audience is meant to hear what is being said. However, it should be noted that sometimes

---

[F]http://cnews.canoe.ca/CNEWS/Crime/2009/08/07/10388781-sun.html

> **WARNING**
>
> Replacing analog devices with digital devices does increase complexity; however, care should be taken that the replacement devices offer actual security. Many devices may use digital, but continue to send the data in the clear where anyone who receives the signal can decode the binary to the data. Many devices do not offer security beyond simple digital encoding. As with any technology, these devices should be investigated and understood before being deployed.
>
> One such example is digital signals from wireless keyboards. Many wireless keyboards operate on the 27 MHz band and while convenient, offer very little in terms of protection if they use encryptions at all. Remote-exploit.org released a paper[G] that detailed vulnerabilities in these systems. Essentially the encryption only applied to keystrokes and the encryption consisted of XORing the keystroke byte with a single byte random value (acting as the key). This only allows 256 possibilities to decrypt, a simple task for even the most low power computers. To demonstrate and audit for this, their proof of concept sniffer only required 20 to 50 keystrokes to get enough data to do a cryptanalysis by testing keys and looking through the decrypted data for common words to verify decryption.
>
> The Remote-exploit.org team went further and designed a dedicated hardware wireless keyboard sniffer that automatically listens for keyboards and decodes keystrokes in real time, writing them to an SD card, to a terminal via USB, an LCD display, or to the Internet over a GPRS or 3G backhaul. The project is called *Keykeriki*[H] and is open source meaning anyone can download the schematics and software and build his or her own sniffer. When one considers the interesting things that are typed on a keyboard through the day, this could be a very valuable tool in the hands of an attacker. The obvious and simple solution is to use a wired keyboard instead.

the microphone is just muted at the sound board but audio is still being sent from the microphone to the base station. This can mean that private, off stage conversation is being broadcast, or more hilariously, private moments not meant to be heard by anyone (such as the infamous scene in *The Naked Gun: From the Files of Police Squad!* where Lt. Drebin forgets to turn off the wireless microphone he is wearing when he goes to use the facilities[I]). Obviously turning off the radio goes a long way to preventing anything being broadcast that you don't want to be. This also has the benefit of extending battery life which is a common problem among these units.

Another option is encrypted microphones. These are relatively new on the market and offer strong encryption (including AES) of the audio from the headset to the base station. These units sometimes even offer a unique key distribution method to discourage static of default keys and usage of long and strong keys. The headset is connected to the base station via a cable (often USB) and the key new pushed to the headset over the wire, removing vulnerability of sending the key over the air and the

---

[G]www.dreamlab.net/files/articles/27_Mhz_keyboard_insecurities.pdf
[H]www.remote-exploit.org/?page_id=187
[I]http://youtube.com/watch?v=N4JBj0kxef8

problem of manually typing a long key on a device without an interface. The major problem associated with these systems is the cost. They are often much more expensive, running higher than 10 times to cost for a headset and matching base station than their unencrypted cousins. They are common in legislatures and government office meeting rooms where security of presentations is paramount, but they are not very common outside of those venues, due to cost. Depending on the information being sent, it may or may not be necessary to use an encrypted microphone, but it is worth considering if the information is sensitive enough.

Analog video signals are fairly easy to secure, mostly by avoiding the low end of the market (anything labeled "spy cam" or sold in those sorts of shops), which are almost always analog. Many brands offer connectivity via Wi-Fi or other digital links which have greater bandwidth as well as higher security than simple analog and leverage existing infrastructure and coverage. The capability to view the video from anywhere on the network is also a benefit as is the ability to digitally record. The usage of a Wi-Fi or other link also means that video can be viewed from anywhere in the world. Many Internet service providers (ISPs) are offering installation of Web cameras to monitor your home when away.

TIP

Care should be taken to secure the network interfaces for these Wi-Fi and network-enabled cameras. Many are set up on networks with public Internet Protocol addresses. This means that unless there is a password or other authentication, anyone can connect and view the video stream. There is even quite the community of people who utilize online search engines to find cameras via phrases used on the Web interfaces. Most often you find a view of a street or lonely hallways, and sometimes you find retail stores or other businesses, but that does not stop people from searching for the elusive "shower cam."

SUMMARY

For the better part of 100 years, we have been using wireless signals to communication. They so permeate the culture that we rarely think about the fact that these devices are tiny radio stations, sending out signals in all directions where anyone with a capable receiver can tune in and pick up those signals. In the world of security, this complacency and lack of understanding can be a great risk when company secrets and private details are being broadcast well outside the control of the user. Rarely is anyone asking questions about the tools we use every day.

Care should be taken by everyone, not just businesses or professionals, to understand the devices we all use and how they operate, and ask questions about the security implications of this device being used in ways not originally conceived. Analog wireless devices are only a small piece of this but pose a great danger due to the ease at which they can be exploited.

Bad Encryption

6

INFORMATION IN THIS CHAPTER

- History
- Proper Encryption
- Passports
- Passport Summary
- Speedpass
- Advanced WPA and WPA2 Cracking

Wireless is a technology with an inherent problem, common to pretty much all standards. It's not really a technical problem, but one of physics. Most wireless gear works on radio frequencies, but others work over light spectrum, such as laser and infrared, and can be considered wireless as well. Most systems have the inherent security problem of a shared medium. Anyone else who is within the range of the signal can receive it just as easily as the intended recipient. For an attacker, this makes sabotage much, much easier. When eavesdropping on these signals became a security issue, the logical and best solution for the problem was to encrypt the signal, or at least the message at the heart of what was being communicated. This is logical since it now means that the intended recipient will know how to decrypt the signal and read the message, but an attacker will just receive an unreadable mess that, without the key, will be indecipherable in a reasonable amount of time.

Given that wireless communications by their nature can be intercepted, protection is very important. Throughout this book, encryption has been mentioned as a way to keep eavesdroppers from reading wireless communications. However, encryption is not an easy thing to implement in products and protocols, and poor implementations can leave a false sense of security and really poor implementations can fail spectacularly.

While it is always a good idea to put crypto into wireless products, it requires a great deal of thought as to how these devices may be attacked and how to do it right. New electronic passports are being distributed to the citizens of many countries, many of which have no idea that the encryption system protecting the digital data on them is fairly weak. The intention was admirable but this actually makes things less secure since the expectation is that the crypto will protect their passport, rather than the citizen taking necessary steps to protect his or her own passport.

Companies are using alternative payment systems to build brand loyalty. Many of these technologies have already been found wanting in their security, putting the reputation and finances of everyone involved at risk.

HISTORY

Messages have always needed to be protected from prying eyes of others for thousands of years. Early forms of this can be seen in the "Caesar shift cipher" so named after Julius Caesar who used it to communicate secretly with his armies. The Caesar cipher is a substitution cipher that takes the alphabet and replaces each letter of the original plain text with a letter of the alphabet a set number of letters down the alphabet. So in a Caesar cipher with a shift of three, the letter A becomes the letter D, B becomes E, and so on. Once the entire message is encoded like this, the message can be sent to the recipient and the process reversed. If the message was intercepted in transit, the message was unintelligible to the interceptor.

Obviously, in modern times, this form of encryption is not considered secure by anyone more than small schoolchildren using it to pass secret messages. However, despite the obvious reasons to not use it, this form of encryption is still in use today in modern systems, more commonly known as ROT encryption. ROT (meaning rotate) is followed by the number of places to shift down the alphabet between the plain text and the encrypted text. The most common of these is ROT13, which means characters are moved 13 places down the alphabet. A becomes N, B becomes O, and so on. This system is also its own inverse; you repeat the same process to decrypt it. To decrypt, you simply move each letter of the encrypted text down 13 letters, looping back to the original letter of the plain text. Among the security and encryption communities, this is considered the most basic form of encryption and should never be used to actually protect something important. It's little more than an exercise in programming and moving bits around. That, however, doesn't stop people from trying to use it.

In 2001, Dimitri Sklyarov,[A] a Russian national, gave a presentation at Defcon 9 in Las Vegas about E-book security. His presentation pointed out that, among other things, one company's portable document format security product that cost about $3,000, simply encrypted the document with ROT13. Adobe was not pleased with this information being made public and shortly after the conference was completed, his hotel room was raided and he was arrested for violations of the Digital Millennium Copyright Act. The case was larger than just the contents of the presentation and included charges based on the fact that the software his employer, Elcomsoft, was

[A]www.eff.org/cases/us-v-elcomsoft-sklyarov

selling were available for download from the United States. Charges were eventually dropped, but the case remains to this day an example of how far companies will go to protect their poor security methods from scrutiny.

A common and recurring theme throughout technology is that encryption solves everything and solves nothing at the same time. Proper usage and implementation is a major factor in how security functions overall. The corollary to this is that as you add complexity to any system, usability and stability rapidly approach zero. Adding too much complexity actually makes systems easier and more prone to breaking. In any system, a balance must be found in order to attain the best security for the situation.

PROPER ENCRYPTION

In the world of encryption, it is worth explaining what a few common terms actually mean as they can differ from their normal everyday uses. The largest of these is about the word "security" and how it relates to protecting something secret from those who should not have it.

Security in the crypto world is measured by how long a message can reasonably remain secure and unreadable in the hands of a determined attacker. If you have a secret, say a muffin recipe, writing it on an index card means anyone can read it. In order to protect it from prying eyes of other bakers, you can encrypt the message in such a way that it is unreadable to someone unless they know the secret (or key) to deciphering it. Now a determined attacking baker who really wants the recipe is willing to spend days or weeks to get it. If you only used ROT13 encryption, the recipe won't stay hidden for long.

The defense used by most encryption systems against a brute force attack is key size. Much like a combination lock, encryption deals with high numbers of combinations to thwart attackers. Brute force attacks involve an attacker performing an exhaustive test of every possible key for the encryption system. An example is the ubiquitous combination lock. Usually this needs three numbers, entered in sequence, to open it. On a lock with 30 numbers, this means that there are $30 \times 30 \times 30$ possibilities of combinations, which equals 27,000 combinations. Assuming 10 s to enter and try each combination, this totals about 6½ h of constant attack to try every combination (though the attacker is likely to discover the correct combination well before that). This is more than an attacker can probably do standing in front of a locker without looking suspicious. Assuming that an attacker can discover the combination on average in about half the key space, security should be built to take that into consideration (guard rotations and changing the combination regularly). In the digital world, the combination is the key and the length of that key specifies the number of combinations possible. These keys can have millions, trillions, or even higher numbers of possibilities, meaning that an attacker has a much larger number to go through in order to decrypt the message. A well-designed algorithm will have a key length and strength sufficient that an attacker would have to work for billions of years in order to exhaust all possibilities of keys to recover the secret. The idea is that the secret will remain a secret for longer than it is useful, usually long after anyone who cares has passed on.

The strength of a well-designed encryption system is commonly thought to be in the secrecy of the key, not the algorithm. That idea is the basis for most modern cryptanalysis. The idea goes that an encrypted message can be obtained by an attacker and they can know everything about the algorithm and how it works and that there are no shortcuts to the plaintext except knowing the proper key. So long as the key remains secure, the message should be as well, even in the hands of an attacker.

A good cryptography system is open for inspection by everyone and has been vetted by experts, and the consensus established that the algorithms used are secure. Many companies and products attempt to add security (or make up for poor security) by hiding the details of their encryption system behind patents, trade secrets, or nondisclosure agreements. This is usually called *security by obscurity* and is generally not considered a very good security system. In security by obscurity, if an attacker learns how the system works, usually he or she is able to figure a way around it or a shortcut through it. In the case of Dimitri Sklyarov, the products being used to secure E-books were hiding behind security by obscurity. Once the secret was revealed, there was no way to resecure the system. It was broken as soon as that information was released. Other companies have to make compromises due to many factors, such as size, cost, and available computing power, when implementing an encryption system, and security by obscurity becomes a necessary part of the system. These situations are sometimes more complicated than others for an attacker, but they all have the problem that once the genie is out, there is no way to put it back in the bottle.

TIP

Anytime you look at products that use encryption, it's worth looking at their descriptions. Terms like *million bit key, totally unbreakable, proprietary algorithm*, or *hacker proof*, are often signs that the company did not use a publicly vetted algorithm and created their own. Further investigation with these companies usually terminates in a statement resembling "trust us," which is a big sign that you might not want to.

A company using the "trust us" defense will often say that their own experts say its secure and that it's been approved by some group or another, but almost always, those people are under nondisclosure agreements and cannot talk about specifics.

PASSPORTS

After the events of September 11, 2001, security of international travel documents was driven to the forefront of everyone's mind (ignoring the fact that the attackers held valid U.S. visas). One of the initiatives was the implementation of additional layers of security within passports through the use of electronics to give the documents some brains as it were. Until that point, a passport was a paper document with many anticounterfeit measures incorporated, but was still just a physical document and a skilled forger could create a reasonable clone of one given enough time and materials. The International Civil Aviation Organization[1] (ICAO) is the international body that oversees commercial travel standards in most countries. The policies and procedures they establish allow planes from pretty much any country to be able to fly to any other country safely and without incident. They are also the main body for establishing travel document standards. They are the group that dictates what information needs to be on a passport, its size, shape, and other features. This allows travelers easy passage between countries and streamlines the process for border control officials.

International pressures led the ICAO to issue DOC 9303, entitled "Machine Readable Travel Documents."[B] This laid out the standards for a machine-readable passport system that integrated digital biometrics and cryptography into passports. The system involved a Radio Frequency Identification (RFID) chip implanted in the passport that could be interrogated by border control, in addition to the physical document. This standard allowed for any country to verify the authenticity of any other country's passport electronic data.

Though some countries choose to include biometrics such as fingerprints into the tag, the standard only requires that the information contained on the photo page of the passport (name, birth date, country, and so on) including the photo be incorporated along with appropriate digital cryptographic signatures of the data. This effort meant that not only would forgers have to modify or recreate the physical document, they would also have to successfully forge the contents of the RFID tag as well and have it match the contents of the physical passport and pass cryptographic muster. Border guards now had an additional tool to verify the identity of the person presenting the passport. The person in front of them should match the photo on the passport, as well as the digital version of the photo on the chip, as well as the two photos matching each other.

However, being that this chapter is about bad encryption, you can guess that the system has a few problems that were made apparent only after these new advanced passports were distributed to the public and they had a chance to evaluate them themselves.

The RFID chip in the passports is a standard ISO 14443 tag, which defines its radio interface to the world. They operate on 13.56 MHz and are powered by the reader like many tags commonly used for inventory control or door access. Unlike inventory control tags, passport tag contents are encrypted so they cannot be directly read from the device until a special 24-digit code, unique to the passport, is sent from the reader letting the passport know that this reader is authorized to receive the information stored on the tag. Once authenticated, the tag sends over its contents along with a digital signature of those contents, signed by the issuing country's private key. The digital signature allows the border control agent to quickly verify that the contents have not been altered and are signed by the country of issue using known public keys from all participating countries.

[B]www2.icao.int/en/mrtd/Pages/default.aspx

In ideal operation, an attacker would be unable to access the contents without the 24-bit key to begin transmission; however, situations are rarely ideal. In order for efficiently moving travelers through passport control, there had to be an easy way to allow border agents to know the key for each passport, short of countries constantly exchanging massive databases of individual codes for every passport issued. The solution was the machine-readable zone (MRZ) on every passport. The MRZ is the jumble of characters at the bottom of the passport photo page. This area is meant to be scanned via optical reader and contains things like name, passport number, date of expiration, and so on. The passport agent swipes the page through an optical reader, saving the agents from having to type all that information themselves into their database. The key is generated from this MRZ data and automatically used to receive the tag contents and display them for comparison to the physical document and the person presenting it. In theory, this means that the tag has a 24-character key, which under ideal circumstances would mean 36^{24} possible keys, a daunting task for an attacker using brute force.

The problem is that since the key is based on set, guessable ranges of values, the key space shrinks dramatically. A birth date in the MRZ, for example, is a two-digit year, two-digit month, and two-digit day. This means that there are only 100 possible years, 12 possible months, and 31 possible days for the key. An exhaustive search of that part of the key would be 365×100, which equals 36,500 possibilities. If you know the rough age of your target, you can narrow that even more to, say, 10 years, 5 years on either side of a guess for only 3,650 possibilities or less. Passports are issued for a limited period of usually 5 to 10 years, depending on the country. This limits the range of the issue/expiration date as part of the key and even more so if RFID tags in passports have only been available for a short time. Passport numbers are a bit trickier to guess; however, some countries have numbering systems based on the date of issue and are sequential. The base value jumps a set amount each day by a fixed amount (passports issued January 1 would be 00001 and increase incrementally, January 2 would start at 10001, and so on). Again, this allows an attacker to narrow down the key space for the passport number.

In 2007, Adam Laurie[C] proved this with a challenge from a reporter. Taking a newly issued passport that had been delivered through the mail, he was able to access the RFID tag and read its contents through the envelope using only information on the envelope and the public records. The outside of the envelope is distinctive as having a passport and includes a postmark and the name and address of the recipient. The postmark greatly narrows the date of expiry to within a few days as well as the likely passport number in some countries. The name and address allowed for a public records search of the target and yielded a birth date, another piece of the key. This means that with a simple reader (since it's a standard interface) and some simple software, a thief could steal a new passport, read all the contents, and put the original back without the victim even knowing that the passport had been read. The information in a passport would allow an identity thief to do all sorts of things. This attack

---

[C]www.theregister.co.uk/2007/03/06/daily_mail_passport_clone/

also does not have to be limited to new passports in envelopes. Since the interface is over the air, an attacker could use a computer in a briefcase and sit next to a target on a train or bus and have a reasonable chance of retrieving the tag contents. Since there is no limit to the amount of times an attacker can try a key (unlike the PIN on your ATM card), it is just a matter of time.

Some countries have identified these problems and attempted to mitigate some of the issues. Some countries have changed the issuing pattern of passport numbers to increase the amount of keys an attacker would have to test. The U.S. implemented perhaps the best-known defense with their passport book covers being layered with a metal foil to act as a Faraday cage to prevent the reader carrier signal from powering the tag and thus sending a signal.

EPIC FAIL

When electronic passports were first being considered, the overwhelming majority of experts thought it was a bad idea from a privacy and security standpoint. Many people questioned the reasoning behind the choice of a radio-based technology when contact-based options (smart cards) were available. In general, RFID is considered a very suspect technology with many conspiracy theories about clandestine monitoring of citizens and the potential for terrorists to use the technology to identify targets by nationality.

Debunking these theories is difficult and was not helped in 2008 at the Hackers On Planet Earth conference in New York City. At that conference, one of the keynote presentations was by Adam Savage, one of the hosts of Mythbusters on Discovery channel. During the Q&A portion of the talk, he was asked by RenderMan about RFID and why the show, well known for debunking urban legends and pseudo-science, had not tackled many of the conspiracy theories surrounding RFID technologies. His answer greatly helped fuel the flames of the conspiracy crowd. He explained that they had investigated the subject for the show, but when the fellow host Tory Belleci participated in a conference call with Texas Instruments (TI) to get some background, also on the line were the chief legal counsels for most of the major credit card companies (who use RFID in some payment systems) – the story continued that the legal counsels all forcefully pressured the show not to show any vulnerabilities in RFID technology. Like many conferences, the presentations were recorded and very quickly a clip of the answer found its way to YouTube.[D]

Shortly afterward, after much media attention, Adam sent out a statement[E] recanting the story as him getting the facts very wrong having heard it from the second or third person. The call did happen; however, not as Adam thought it did. Host Grant Imahara was the team member on the call and there was only one legal representative attending. At no time did anyone pressure the show to not pursue the angle (Texas Instruments was trying to arrange for it to happen) and it was the show's decision not to go forward with the tests.

So the epic fail and the lessons learned are twofold. First, asking a simple question about the security of a system is a good thing but the answer may be more than you bargained for. Second, particularly with security technology, be very aware of the facts before making a statement about its positive or negative impacts as many people are often dependant on these technologies for their safety and security and will fight to set the matter straight if you are wrong in any way, leaving you looking foolish.

[D]www.youtube.com/watch?v=-St_ltH90Oc
[E]http://news.cnet.com/8301-13772_3-10031601-52.html

The cryptographic problems get worse once you consider that not all passports are created alike. While interoperability was a goal with the ICAO standards, not all portions of it were mandatory and countries were allowed to make their own decisions in some cases regarding the authentication methods. One method available against cloning one chip to another is called active authentication (AA). With AA, the legit passport is embedded with a private key that cannot be directly read but can be used cryptographically with a public key on the tag to verify those tag contents are original to that tag. A cloned tag would not have that private key and thus would fail the test. This feature is optional, however, and many countries have chosen not to support it, meaning systems have to deal with it not being there. Researchers at The Hackers Choice[F] found that if they copied the tag contents, altered them, and deleted the original public key on the tag, they could sign the tag contents with their own key from a nonexistent country and pass several automated passport control systems (though doubtful any that included a guard). The problem was that many of these systems checked to see that the contents were signed by the key pair on the tag, but failed to check if those keys were trusted and actually came from the country it claimed to be issued from. Many countries, systems do not do this and would need to patch or upgrade their systems to perform this check.

PASSPORT SUMMARY

While the battle between document issuers and document forgers is a never-ending battle, the addition of digital components makes the job just that much harder. However, no system is completely safe, and compromises made for ease of implementation

---

WARNING

While electronic passports are here to stay for the foreseeable future, many people are uncomfortable with them. There are many who advocate damaging the RFID tag to prevent any reading. While this may sound like a good idea, it is worth becoming very familiar with the implications of this for your country.

In the United States, an electronic passport where the tag fails is still a valid document[G] for travel until its expiration. The U.S. border agents may give you more scrutiny than other travelers, but you will not be stopped solely for a broken tag since it would be very bad to leave travelers stranded at the border if they sat on their passport accidentally. What is not clear is if other countries apply the same consideration.

In some circumstances, it may be also considered tampering with a federal document if you intentionally damage the tag, which may have unforeseen consequences when attempting to use it to cross a border.

While protecting your personal security and privacy is important, it should be done in accordance with your local laws and the laws and policies of your destination.

[F]www.thc.org
[G]http://travel.state.gov/passport/eppt/eppt_2788.html#Fourteen

and use severely lowered the amount of security available in passports. The lack of understanding in both government and private citizens about these weaknesses mean that very often, they will grossly overestimate how secure a document is with digital components, calling it *uncloneable* and *forgery proof*. Terms and ideas like that make people much more likely to put more trust into such documents and likely to let reasonable forgeries pass with less scrutiny that their nondigital counterparts, thus defeating the whole purpose of the system in the first place.

SPEEDPASS

Payment systems have an obvious need for encryption. Given the financial incentive to attack these systems, it's a natural fit to encrypt payment information whenever possible. Payment systems also demand ease of use and quick turnaround to be appealing for customer, retailer, and payment processor alike. To that end, many companies have instituted their own payment options for their stores and services over the years, but in 1997, things took a leap when Mobil Oil Corp. created the Speedpass system.

The Speedpass was a small plastic fob on a keychain that contained a glass capsule 134.2-MHz digital signal transponder (DST) from Texas Instruments. These tags were small and cheap to purchase in quantity, allowing them to be given away for free. The tag's identifier was tied to a credit card or checking account of the customer's choice and automatically charged the account with just a wave of the tag past the reader built into the gas pump. The theory was that it was more convenient since all you needed to purchase fuel was the tag – removing the need to deal with cash or credit cards inside, you could instantly pay at the pump. The tag also played into brand loyalty as the convenience of it and the exclusiveness to one brand of service station meant that customers were more likely to return to those stations. The convenience also led to an increase in sales at the convenience stores attached as people were not handling real money and thus were apt to spend more.

In 1999, Mobil Oil Corp. was purchased by Exxon, which became Exxon-Mobil and expanded the program out to the Exxon service stations as well making it one of the first large-scale RFID deployments used by the general public. Similar technology has been rolled out for other customer loyalty cards and payment systems, but the Speedpass was one of the most notable in terms of size (millions of tags and thousands of locations) and also in failure, as it was the first major system to fall under external research.

The tag operated with a challenge/response authentication system. The interrogating reader sends a 40-bit challenge to the tag, which the tag then uses to calculate a 24-bit response, using a cryptographic algorithm incorporating a unique 40-bit key stored on the tag. The tag sends that response along with a unique 24-bit identifier, the unit's serial number. The reader computes its own response to the challenge using the expected key for that tags serial number. If the readers expected response matches what the tag sent, the tag has confirmed that the tag as authentic and proceeds to debit the credit card or account stored in the backend database that is tied to that unique tag. All of this takes less than a second at the gas pump to process and allow for operation. The system also allows for multiple tags to be attached to the same account, allowing families and businesses to use them for easy purchases, a very handy feature for fleets and families.

For many years, there was interest in the security of the system, but the incredibly tiny transponders were a mystery to researchers. The cryptographic system in use was proprietary and shielded behind nondisclosure agreements and programming application program interfaces. It was the proverbial security by obscurity. Without knowing the internals, an attacker could only see what went into the tag and what came out, but had no idea of the intermediary process and no idea how to recreate it.

Security by obscurity by definition only works so long as the secret remains secret; however, once that secret is known, the system that depends on it falls apart. Often, all it takes is a matter of time and effort. For Speedpass, that time and effort was put in by a team from Johns Hopkins University in 2005.[H]

The Johns Hopkins team obtained an evaluation kit and a number of sample Speedpass tags. The key piece of their research was a presentation from a Texas Instruments (TI) engineer at the AES4 Conference in May, 2004, in Bonn, Germany.[I] In these slides, the engineer gives a rough block diagram of the internals of the algorithm. This diagram was lacking many important details, but it provided a starting point for a "black box" approach to reverse engineering the algorithm. The team used their own DST tags, identical in function to the Speedpass tags, but they were able to specify their own key that allowed them to follow the known constant input and compare it to the output and use that to reconstruct the missing pieces of the algorithm internals. From there, they also changed individual bits of the input key to see what changed on the output to understand further details.

The team ended up with a working model of the algorithm, written in software that provided the same resulting output from the known input provided to a real DST tag. This meant that they had confirmed a working simulator for known keys. But what about unknown keys, like on the Speedpass?

The devil is in the details of any algorithm and the technology it is implemented on. RFID tags are very low-power devices, and since the DST tags are externally powered by the reader, they don't have much power to work with in order to do the necessary computations of a complex algorithm, and most modern cipher algorithms are CPU intensive. To get a solution that fit on the hardware at hand, the TI engineers designed their own algorithm instead of using a publicly trusted one. They also chose the comparatively small key size of 40 bits to find the balance between the best security and the best performance on the tag. These decisions were not wrong by any stretch; it's

---

[H]www.usenix.org/events/sec05/tech/bono/bono.pdf
[I]Originally at www.aes4.org/english/events/aes4/downloads/A. ES_UICE_slides.pdf site has since expired. Alternate download at www.renderlab.net/files/AES_UICE_slides.pdf

an issue of how long the security of the system would be in place before it could be upgraded, replaced, or phased out. In this case, it ended up lasting less than 10 years before it was first publicly compromised.

The research team exploited this weak key, applying computing power that was previously unavailable when the system first came out. They knew that the 40-bit key provided 1,099,511,627,776 combinations, which may seem like a lot, but its well within the range of a computer to calculate within a reasonable time frame.

NOTE

The term *reasonable* when talking about cryptography or cryptographers can have a greatly different meaning from that of the general populous. In cryptography, where an exhaustive search is expected to take billions of years, finding a way to drop that to a few hundred is quite a feat and considered a lot more reasonable, especially when advances in computing power may shave that value down even further.

The researchers knew that they could capture sample output from an unknown tag by simply using their own reader to interrogate it with a known challenge. In this way, they know the algorithm, the challenge, and they can observe the response. The only factor left was the secret 40-bit key. The easiest way to recover that key (with the intention of duplicating it later) was to capture a couple of exchanges and then test every possible key to see if it obtained the same output given the same input at the tag being tested.

The team found that a standard 3.8-GHz single-core PC could compute about 200,000 keys per second, which would take a while to accomplish an exhaustive search (about 64 days). To speed this up, they implemented the cracker program in hardware using field-programmable gate arrays (FPGAs) and small-programmable processors that are generally much, much faster than general-purpose CPUs at performing single tasks. Their implementation allowed 16 million keys to be tested per second on a single FPGA, which dropped the timetable down to about 21 h. They were able to use this to recover the key on a previously unknown Speedpass. Spreading the workload across 16 cards, however, allowed them to drop the time to around 1 h (on average).

To verify their work and prove themselves to Texas Instruments, the team had TI provide five tags with keys selected and known by TI. Within 2 h, the team recovered the keys to all the provided tags with their parallel cracking setup, thus confirming its abilities.

Taking the research a step further, the team applied the time–space trade-off and built a "rainbow table" that contained precomputed arrays of inputs, outputs, and secret keys. At around 10 Gb, the table could be used to simply look up the key of a tag based on the output from a known challenge, reducing the time to recover a key to about a minute, or as fast as a computer could do a search for the right challenge/response.

Applying the attack, the team built the next logical thing, a tag simulator, using their software implementation of the tag to emulate a legitimate tag in the field and use it to make a purchase. The necessary equipment took up most of the back seat of the test vehicle, but they succeeded in obtaining a real tag's secret key and then emulating it to commit a "fraudulent" purchase (the tag was owned by a researcher, so it wasn't fraud, but an attacker would not be so discerning). The equipment size is just a matter of engineering to reduce to easily an easier, more portable size to allow for sneak attacks.

The resulting fallout from this research was not mass fraud or a breakdown of the entire payment system. The team responsibly kept back details of the algorithm and did not make their implementations public. This made it harder for any attacker to copy their work but allowed them to prove that the research was correct. In addition, Exxon-Mobil was already beginning plans to phase out the older tags in favor of newer, more secure tags.

The conclusion best drawn is that the air interface of the Speedpass allows an attacker access to the communications. The security of those communications depends entirely on the strength of the algorithm protecting it, and that what was considered secure one day, may not be secure the next day.

ADVANCED WPA AND WPA2 CRACKING

Chapter 1, 802.11 Wireless – Infrastructure Attacks focused on Wi-Fi infrastructure attacks and included weaknesses in Wi-Fi Protected Acess (WPA) and WPA2. It is worth exploring the future of these attack vectors and the security that WPA Pre-Shared Key (WPA-PSK) and WPA2-PSK provide.

Current attacks on WPA and WPA2 in PSK mode depend on the guess-ability (or entire lack) of the passphrase used. While an exhaustive search is still out of the realm of possibility, if an attacker can guess your passphrase or you use a very common one, there is a good chance that you may be compromised. All of this depends on the ability to choose a passphrase that is longer and more random than the ability of an attacker to muster enough computing power to end up testing that passphrase in a reasonable amount of time. The idea was that a sufficiently long and random passphrase was enough to thwart most attackers and that the SSID being used as a salt made those attacks even more difficult.

In late 2009, Moxie Marlinspike launched wpacracker.com and the wireless world took notice of the availability of cheap processing power and its potential in password cracking. The service uses Amazon's EC2 cloud computing service to allocate processing power for cheap as compared to actually building it yourself. For $35, the service will test a capture (in the same way coWPAtty does) for a single SSID against 135 million word dictionary and return results in about 20 min (no guarantee of a success though). There are also options for a dictionary in German, as well as an extended English dictionary. The final one-off option is all the dictionaries combined for a total of 520 million words against the target network.

One can imagine that a determined attacker could probably muster more than $35 to devote to having half billion words tested in a very small amount of time.

The 802.11i standard allows for a minimum of 8 and a maximum of 64 characters with most users falling well below the 20-character mark. Chances are statistically significant that the passphrase for many potential targets is in that list and security could be compromised in less than an hour for less than a console video game. When you consider that attackers don't even have to use the automated service, they could just operate their own virtual cracking machine in the cloud where money is the only limit or just hire or create a botnet to do something similar at their beck and call.

This availability of cheap processing power should make anyone reconsider "good enough" in their security solutions. In the case of WPA-PSK and WPA2-PSK, the power available to an attacker should make one pause to think if they are capable of coming up with a suitable passphrase or if it's time to evaluate other solutions like WPA2-Enterprise, which uses external authentication and is not subject to the same brute force attack vectors.

SUMMARY

While encryption is a solution to protect wireless communications, it's not simply a matter of bolting it onto an existing product and calling it a day. The selection of algorithm, the implementation, and a reasonable expectation of the length of time something will last should be considered.

Algorithms used should be publicly vetted and trusted by the cryptanalysis community. Any companies with proprietary or homemade algorithms, while they do work, should be considered with a grain of salt as you cannot independently confirm the security in place. It's a matter of "trust us" from the developer.

Even if a trusted algorithm is used, poor implementation can cause security to suffer greatly if the keys are easily discovered or guessed. As well, understanding what you are protecting can help you evaluate if the solution will protect your secrets for longer than they will be useful.

Crypto is necessary in our daily lives to keep things private and secret. It's also really hard to do right. Care should be taken to evaluate any products for their security implications, even those that seem to take it to heart.

For any products that use encryption, there are several guidelines that you should keep in mind when evaluating their potential effectiveness:

- Algorithms should be freely available for scrutiny and analysis.
- Algorithms should have a history of withstanding rigorous attacks.
- Anything secret, proprietary, or that contains "miracle" marketing claims should be treated with skepticism.
- Assume that the attacker knows the algorithm. This should not help them in any meaningful way.

- Acknowledge that any system can be broken with enough time and resources.
- Scale those resources appropriate to the length of time something needs to remain secret.
- Encryption should only be part of a larger system of security.

Endnote

1. www.icao.int; [accessed 12.09].

Cell Phones, Personal Digital Assistants, and Other Hybrid Devices

INFORMATION IN THIS CHAPTER

- Hybrid Devices
- History
- Anatomy of the Attack
- The Attacks
- Future Attacks
- iPhone Summary

So many devices on the market today have their roots in the visions of the future from the past. Mankind has always looked to the future with wonder and hope. Television shows like *Star Trek* and just about every other science fiction show of the 1950s and 1960s showed a future full of gadgets to make our lives easier and have knowledge at our fingertips.

There's no mistaking that the cell phone's design was driven by the communicators in *Star Trek*, and that devices like the Amazon Kindle have taken on a role resembling that of *The Hitchhiker's Guide to the Galaxy*, though without the helpful "Don't Panic" inscription (though one user corrected that omission: http://seanreiser.com/content/getting-kindle-etched). One aspect of these devices and the predicted future is the security of these devices and the risks that they pose to things people generally hold very dear. Things like privacy and integrity of the information retrieved and stored on those devices are something rarely discussed on those shows, and only now are these challenges becoming apparent as issues we have to deal with – something our past vision of the future never prepared us for.

HYBRID DEVICES

Hybrid device is a term that means different things to different people. Here, they are best defined as those devices that do more than just what their purpose is on the surface. An example is a modern cell phone. In addition to making and receiving calls over traditional cellular systems like global system for mobile communications (GSM) or code division multiple access (CDMA), they also have a host of other features, such as Wi-Fi and Bluetooth connectivity, and the ability to access the Internet over cellular data networks, such as Enhanced Data Rates for GSM Evolution (EDGE) or 3G.

These devices pose a new set of problems for administrators and users since they are capable of doing far more than what they were primarily purchased for, and risks may not be apparent.

One recent example of this was a rash of attacks against Apple iPhones in late 2009.[A] These attacks targeted phones that had been "jailbroken," an action taken by phone owners to run unsigned programs and content on their phones. Many users suddenly found themselves locked out of their phones and their lock background changed to a message asking for money to be sent to a PayPal account in exchange for tips on how to secure their phone. Later, a worm[B] was discovered in Australia that propagated through jailbroken phones that RickRolled the phone by changing the background to a picture of Rick Astley and changed the ringtone to that of "Never Gonna Give You Up," the song of RickRolling fame (or infamy).

This fairly harmless prank caught the world's attention to the risk of hybrid devices and made them realize that they are more general-purpose computers than single-task devices. The threats need to be viewed in context of general computing and not as cell phones or whatever they are being sold as.

HISTORY

People in modern society carry a surprising amount of data with them, even if they don't understand or realize that they do. For many years, it was the leather organizer stuffed with notes and business cards and calendars. After that, the organizer was supplemented by the cell phone and the ability to always be in contact. Once cell phones hit the mass market, they became indispensable. These early phones did little more than just one function, to send and receive calls. Even just a few years ago, a business traveler was hard pressed to be separated from his or her monolithic-use cell phone. After a while, computing advanced and the organizer was replaced by a personal digital assistant (PDA). Just a few years ago, the Palm Pilot was the ubiquitous device among the business set. The amount of information that one could now carry far outstripped the previous capability for any one person to carry with them, let along in a pocket. This suddenly started presenting a more interesting target for attackers. The amount of information present and the often sensitive nature was tempting. These devices, however, had very limited communications options, and short of assaulting the target and stealing the Palm Pilot, the options for an attacker were limited as to how to obtain information off of targeted devices. If you are jumping the target anyway, you may as well get the information from them directly.

[A]http://news.bbc.co.uk/2/hi/technology/8380265.stm
[B]www.sophos.com/blogs/gc/g/2009/11/08/iphone-worm-discovered-wallpaper-rick-astley-photo

In the early 2000s, the Blackberry took the world by storm as well as other devices that combined the functions of a cell phone and a PDA in one device making for a more convenient device to carry around. This new generation of devices introduced new and interesting communication options not only for the users but also for the attackers.

Today's communications devices are interesting from an attacker's perspective. Consider a modern Apple iPhone. These popular phones have surprisingly little resemblance to the phones of just a few years ago. Previously, phones did one thing and one thing only; they made and received calls. They did that very well and became indispensable to business people and soccer moms alike. However as Apple's marketing points out, there's little that the iPhone can't seem to do. These devices are basically portable computers that include making and receiving phone calls as one feature rather than as a dominant purpose.

These devices also exist outside the control of most information technology administration functions. Their highly portable nature has them outside the watchful eye of any infrastructure setup to protect company data or assets secure inside the workplace. Also, these sorts of devices are very popular with nontechnical users who are unaware of the risks of using these devices on insecure networks in public places. A recent test by Renderlab.net found that, at a large shopping mall, the assembled mass of several thousand people had at least 247 devices among them with Wi-Fi capability turned on and were attempting to communicate with the Internet. Logins and passwords for social networking sites like Twitter and Facebook were plentiful and the opportunities to inject data to cause chaos were quite rampant. Walking into any busy public with simple tools can reveal hundreds of mobile phones, each one potentially vulnerable.

ANATOMY OF THE ATTACK

Late 2009 was an interesting time for Apple and iPhone security in general. The iPhone gained more market share, and new models added new and interesting features for users to play with. The App Store installed on every phone allows users to add software and customize their phones capabilities. New productivity applications, food selection, and even applications remembering where you parked are all available.

A common complaint about the iPhone is that it is a closed system. The iPhone is only capable of running software that is purchased and downloaded through the App Store, and all of these are subject to review and approval from Apple before they are included. Some applications are free while others cost as little as $0.99 compared to others costing several hundred dollars (though not many do). Apple has chosen to restrict what applications can be accessed in terms of the operating system (OS) and other core functionalities, such as stored music. The process of approval for an application is heavily veiled and mysterious to developers who often complain about not knowing why their application was denied.

Apple's reasoning behind these restrictions range from copyright and antipiracy requirements, to stability and performance reasons. Critics counter that Apple is trying to maintain a stranglehold on revenue streams and limiting competition to those streams.

EPIC FAIL

Apple, Google, and AT&T got into a fairly large mess with one another in July of 2009.[c] Google had submitted an application for its Google Voice service, a service that allowed users to have a single "Google phone number," which could be routed to multiple phones simultaneously as well as central voice mail and text messaging from multiple devices. The calls would be routed over the Internet using Voice over Internet Protocol (VoIP) instead of normal telecom lines, heavily reducing costs to the end user. Obviously, this would be a useful service for anyone with multiple phone numbers and locations to worry about, and anyone wanting to save money by using VoIP for normally expensive long distance calls.

Controversy sprung up when it was revealed that Apple had rejected the Google Voice application from the store and purged all third-party applications that provided an interface to the service. A subsequent Federal Communications Commission investigation made the situation all the more strange since replies from each company painted different stories. Google claims Apple outright rejected the application; Apple claims it was still reviewing the application; and AT&T said it had nothing to do with the rejection despite its obvious benefit of not seeing such a service exist, and older agreements with Apple not to allow some VoIP functionality. AT&T's insistence that they had no part in the situation was despite the potential of reducing their existence to a "dumb pipe" that others route call traffic over, obviously restricting their ability to offer value-added services and cutting them out of the revenue stream.

In October 2009, AT&T announced that it would allow VoIP functionality over its 3G connection. Products like Skype previously worked over the Wi-Fi connection but were restricted from working over the 3G link. That appears to have changed with this announcement. While no specific announcement was made in regards to Google Voice specifically, it does seem to indicate that the potential exists for it to be approved.

The controversy is ongoing and certainly not going to be the last, but points to a larger epic fail in that hybrid devices are everything and nothing at the same time. To AT&T, the iPhone is a mobile phone. To Apple, it is a sales platform for music and applications. To Google, it is a sales platform and a computer platform portal to its services. Each company is going to treat it differently, and unfortunately, the consumers are caught in the middle of these disputes, and in some cases will not be able to benefit from the available technology due to these disputes.

While the AT&T issue seems to have settled and VoIP is allowed, this same problem is due to creep up again with other providers offering the iPhone and customers wanting to have similar functionality.

Jailbreaking

Because of this closed nature, since moments after the first iPhones were released, users had been seeking ways to bypass the restrictions placed on what software could be run and from what sources it could be obtained from. These culminated in what became known as *jailbreaking*. Jailbreaking means to break the digital locks Apple has put in place and load applications not approved by Apple. This could allow other applications to be loaded (an alternate App Store) or for the phone to be "unlocked," which means to remove and restrictions put in place to lock the handset to a specific carrier, allowing its use on other networks.

---

[c]http://gizmodo.com/5324268/apple-rejects-official-google-voice-iphone-app

> **WARNING**
>
> Jailbreaking an iPhone or modifying any manufacturer's handset beyond methods they allow (ringtones, wallpapers, and so on) will most likely void any sort of warranty on the device in the eyes of the manufacturer and the carrier, and may introduce unforeseen problems. The terms of service with the carrier usually spell this out quite clearly and may result in termination of service and loss of warranty on the handset.
>
> So if you jailbreak your iPhone, and it gets damaged, you are on your own for repair/ replacement costs, and considering the nonsubsidized costs of these devices, that can be an expensive proposition.

Initial unlocks required fairly extensive modification (including hardware) to perform, but it has been reduced now to a simple point-and-click interface. In the end, these applications load an alternate App Store application (in addition to the Apple offering) that bootstraps the ability to load further, nonapproved software. The most popular of these packages is the PwnageTool from http://blog.iPhone-dev.org, which loads the Cydia App Store application. Once loaded, the Cydia store allows users to load applications that are able to do things beyond what Apple permits through their App Store, including much deeper access to the file system and the OS, which not only allows more functionality but also more potential for damage to critical system files.

While the jailbreaking process does not specifically introduce any vulnerability, it does however introduce some complexity that users may not understand. This is what was being exploited by the recent iPhone worms.

As part of the Cydia App Store, there is an option to install and enable an OpenSSH server on the phone itself since some applications available on Cydia are command line only and the easiest way to connect is through a local Secure Shell (SSH) connection. This is not a major feat for the iPhone, however, since its OS is UNIX based it is not available from the default Apple firmware.

The cause of vulnerability was not in that owners had jailbroken their phones, but that they were not understanding what they were doing in the process. In this case, many owners installed an OpenSSH sever onto the phone without understanding the consequences. In this case since the iPhone has a UNIX-based OS, it has users built in: root and mobile. Root is the super user of the system and can do anything. The user "mobile" is a restricted user that most applications run as and the owner of most of the interesting user-generated files like the e-mail database, calendars, short message services (SMS), and so on. Both of these accounts use the same default password of "alpine," which is widely documented around the Internet, along with large notes in the Cydia documentation to change those passwords if you enable the SSH server.

Users being users, though, many did not change the passwords and this led to a new twist on a very old attack vector.

Despite the media rhetoric and noise about attacks on jailbroken iPhones through SSH, this is far from a new attack. It is probably one of the oldest attack vectors known: a user leaving a default password on an administrative account.

While it may be new and interesting that this applies to a mobile phone, these devices are now closer to mobile computers than they are mobile phones and should be treated as such. Users, however, have not quite caught up to this line of thinking. To many of them, they are still simple phones first and not general computing devices.

A manual version of the attack over an existing wireless network is fairly easy to do. The only difficulty is that devices like iPhones are usually used on demand and can come and go from the network as they are used and put into standby. Airports, coffee shops, libraries, and any other locations with free Wi-Fi are likely places to find iPhones and iPods using the wireless network.

The first step is identification. Since iPhones are network devices, they require an Internet Protocol (IP) address to work over Wi-Fi just like any other device on the network. A simple Nmap or other port scan will identify devices on the network and report what services are running. iPhones will typically have port 62078 open, which is iPhone sync, a UPnP service for the iPhone to share content with other Apple devices. This single open port is usually a good indication, but also confirming the media access control address corresponds to those used by Apple. Jailbroken phones with SSH enabled will have port 22 open for SSH (the default port for the SSH service). A simple Nmap output is shown in Figure 7.1.

Once an iPhone has been identified, it is simply a matter of logging in over SSH as root to the phone with the default password of "alpine." Figure 7.2 shows logging in to a jailbroken iPhone and the root folder.

Once logged in, it is a standard UNIX command prompt. The most basic commands, "ls," "cp," and "rm," are available along with things like Secure File Transfer Protocol. While not every common UNIX command may be available, there is plenty enough to move data on and off of the device.

NOTE

During the whole alarm at the initial rash of automated attacks, most news sites were trumpeting the same line about changing the root password on the iPhone to keep out invaders. What many sites failed to mention was that the username "mobile" had the same password of "alpine" and had access to most of the same interesting things that an attacker would be seeking on a mobile device (contacts, mailbox contents, SMSes, photos, and so on).

These articles were probably doing more of a disservice to their readers by failing to mention the user "mobile" requiring a password change as well, since a well-informed attacker would try out user: mobile, password: alpine.

Once in as root, there's really no limit to what you can do on the system. It's just like any other UNIX-based computer system. The most likely of these things would be to steal the personal data from the iPhone's owner. The more interesting files on an iPhone are located in the /private/var/mobile directory.

The calendar application stores its data in the /private/var/mobile/Library/Calendar directory as shown in Figure 7.3.

The mail folder which stores local copies of e-mails received is stored in /private/var/mobile/Library/Mail with each account having its own mailbox. Figure 7.4 shows an example of a mail folder.

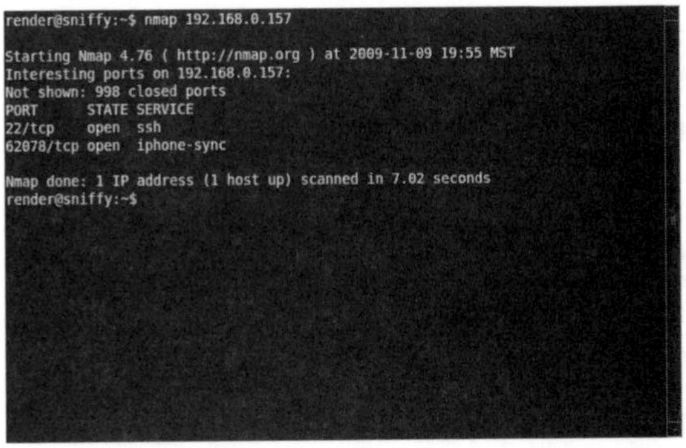

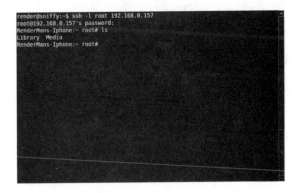

```
render@sniffy:~$ nmap 192.168.0.157

Starting Nmap 4.76 ( http://nmap.org ) at 2009-11-09 19:55 MST
Interesting ports on 192.168.0.157:
Not shown: 998 closed ports
PORT     STATE SERVICE
22/tcp   open  ssh
62078/tcp open  iphone-sync

Nmap done: 1 IP address (1 host up) scanned in 7.02 seconds
render@sniffy:~$
```

FIGURE 7.1

Sample Nmap Output of a Jailbroken iPhone

```
render@sniffy:~$ ssh -l root 192.168.0.157
root@192.168.0.157's password:
RenderMans-Iphone:~ root# ls
Library  Media
RenderMans-Iphone:~ root#
```

FIGURE 7.2

Logging into an iPhone as Root

```
RenderMans-Iphone:/private/var/mobile/Library/Calendar root# ls
Calendar.sqlitedb
RenderMans-Iphone:/private/var/mobile/Library/Calendar root#
```

FIGURE 7.3

iPhone Calendar Data File

```
RenderMans-Iphone:/private/var/mobile/Library/Mail root# ls
AutoFetchEnabled                        IMAP-renderman@www.renderlab.net
Envelope Index                          MFData
IMAP-renderlab@gmail.com@imap.gmail.com Mailboxes
IMAP-renderman@mail.c2security.org      metadata.plist
RenderMans-Iphone:/private/var/mobile/Library/Mail root#
```

FIGURE 7.4

iPhone Mail Folder

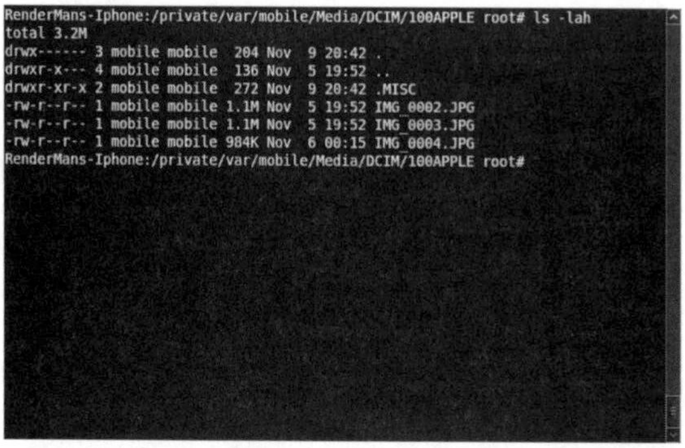

```
RenderMans-Iphone:/private/var/mobile/Media/DCIM/100APPLE root# ls -lah
total 3.2M
drwx------ 3 mobile mobile  204 Nov  9 20:42 .
drwxr-x--- 4 mobile mobile  136 Nov  5 19:52 ..
drwxr-xr-x 2 mobile mobile  272 Nov  9 20:42 .MISC
-rw-r--r-- 1 mobile mobile 1.1M Nov  5 19:52 IMG_0002.JPG
-rw-r--r-- 1 mobile mobile 1.1M Nov  5 19:52 IMG_0003.JPG
-rw-r--r-- 1 mobile mobile 984K Nov  6 00:15 IMG_0004.JPG
RenderMans-Iphone:/private/var/mobile/Media/DCIM/100APPLE root#
```

FIGURE 7.5

iPhone Photo Directories

Photos are probably one of the more interesting targets, particularly as potential blackmail material or as a curiosity. Photos taken with the phone can be found in the /private/var/mobile/Media/DCIM folder, usually under a directory named 100APPLE or another similarly named directory. Files are simple JPEGs and can be viewed with most image viewers, and as Figure 7.5 shows, they are owned by user mobile, thus reenforcing that changing the password on the mobile user is important as well.

At this point, an attacker can download a great deal of private content; however, the attack relies on a persistent connection and can take time to download all interesting content. Obviously much can be scripted to speed things up, but there is also the other problem that our target may not be connected to the local network at the airport, hotel, or wherever he or she is being attacked.

iPhones have an interesting behavior. As noted in Chapter 2, 802.11 Wireless – Client Attacks, some devices are very persistent in advertising for and about connecting to networks in their preferred list. iPhones in particular seem to have been designed to do this almost forcefully.

Within the Wi-Fi networking menu of the iPhone, there is a note at the bottom of the menu that states "Known networks will be joined automatically. If no known networks are available, you will be asked before joining a new network." Since these are hybrid devices, they have the ability to connect to the Internet over a 3G connection as well. This obviously will incur costs to the user and as a courtesy, the OS is designed to use the 3G as a connection of last resort. Simply put, the system will check if a Wi-Fi connection that it has been connected to before is present. If it is, it will automatically join it and use that network. If one does not exist, then if there are any around, the user is offered the opportunity to connect to one of those networks. If none exist or the user cannot or declines to connect, then the 3G connection is used. All of this is provided that the Wi-Fi connection is enabled in the first place.

This creates the interesting situation for an attacker. Since the iPhone favors Wi-Fi connections and is constantly searching for home networks and will automatically connect without user prompt, an attacker can easily take advantage of this.

Using tools from Chapter 2, 802.11 Wireless – Client Attacks, such as KarMetasploit,[D] it is trivial to setup a fake access point that responds to any probe request and emulates the networks on the preferred list of the iPhone. Once the iPhone connects, it is silently given an IP address and now the attack has a direct connection to the device with no user interaction. The wireless community has found that the iPhone is one of the easiest devices to exploit this behavior on.

The problem of intermittent connections due to the device being put in standby is strangely solved by the environment the attack takes place under. Most iPhone users are not using their devices constantly. They use them for short periods and then place them in standby. There are places, however, where constant usage is almost guaranteed, typically in any place where someone has to wait long periods of time. One of the easiest places to exploit this is on an airplane. Users will typically disable the 3G functionality of the phone before the flight as per Federal Aviation Administration regulations but often not disable Wi-Fi. Several miles in the air, most users will be listening to music or watching movies on their devices, blissfully unaware that their device is still seeking networks to connect to. All an attacker has to do is be in range, which is easy on a plane, and then start up the attack. Automated systems can make this take only a few seconds, leaving more than enough time for it to work in other situations, like two cars traveling the same direction or even while stopped at a red light.

TIP

While most times, an attacker would have to be very determined to get on the same flight as a target, there exist targets of opportunity. A very bored person in the back of economy class can wreak a great deal of havoc to keep themselves entertained and it is highly unlikely that they would be noticed until well after the flight and they had parted ways.

This also applies later to automated viral and worm attacks. Continuous contact in a contained area increases the likelihood of an infection spreading to your device, particularly with many airlines now offering onboard Wi-Fi capabilities.

[D]www.metasploit.com/redmine/projects/framework/wiki/Karmetasploit

THE ATTACKS

In early November 2009, news sites and technology blogs[E] were all abuzz with news of an attack on Apple iPhones. Several Dutch users of jailbroken iPhones reported that their devices had been attacked and a ominous message left on their lock screen wallpaper indicating a Web site to visit to find out how to secure their devices. The Web site demanded five euros for instructions on how to secure the phones. After news broke, the attacker very quickly removed the demand for payment and posted instructions on how to secure the phones as well as an apology to users he had attacked. As guessed, his preferred method was default root passwords on jailbroken phones. This was the first public demonstration of the capacity for this vulnerability, but certainly not the first time it had been quietly abused.

Later in November, another iPhone attack appeared in Australia[F] utilizing the same attack vector. Except this time, the attack was automated and could spread automatically from phone to phone via Wi-Fi networks that vulnerable devices shared. The Ikee worm, as it was known, was the brainchild of Ashley Towns, a 21-year-old student. The worm infected jailbroken iPhones through default SSH passwords, but this time changed the background image to that of Rick Astley and the ringtone to that of the song "Never Gonna Give You Up" of RickRoll fame. The worm would then attempt to upload itself and do the same to other phones on networks joined by the infected device. Reports only indicate 100 phones or less were infected, but considering the small percentage of iPhones that are jail-broken, this number is interesting. While it is unclear if the actions of the Dutch extortion attempt had anything to do with inspiring the Ikee worm, it did bring its author to the attention of iPhone application developer Mogeneration who hired Towns as a developer after an interview process after hearing about his worm. They claim it was not due to the worm and that he passed their iPhone developer test with flying colors as any other employee would have to. There is great debate if hiring him is rewarding negative behavior and sets a dangerous precedent.[G] Some see it as positive reinforcement of a negative action and that if there is the hint of a reward, others will attempt to copy that behavior in hopes being rewarded. The end result may be a great many people attempting to write worms for the iPhone or other devices in hopes of landing a job. Some see it as asking the fox to guard the hen house, hiring

---

NOTE

It is still unclear as to whether or not the worm could spread via 3G connections. Many providers differ in how devices are presented to the Internet at large. Many providers of 3G gateways heavily fire wall the incoming connections for the IP space used by wireless devices. They also may deny interclient communication. While this prevents some attack vectors, some providers may not be as restrictive and allow the types of connections required to spread this type of malware. Much of it depends on the infrastructure of the carrier that the device is connecting to and can vary from provider to provider.

[E]www.sophos.com/blogs/gc/g/2009/11/03/hacked-iphones-held-hostage-5-euros/
[F]www.sophos.com/blogs/gc/g/2009/11/09/worm-author-tells-media-initially-infected-100-iphones/
[G]www.sophos.com/pressoffice/news/articles/2009/11/iphone-worm-job.html

the best talent possible. It is certainly not a debate that will end soon, nor likely ever to be resolved.

These first two attacks were fairly benign and did little to no damage to the attacked devices. However, they defiantly showed the world that there were a significant number of jailbroken devices that could be compromised easily. It also showed that the user base was fairly un-savvy about the risks to these hybrid devices.

Shortly after the first two attacks, a third worm showed up, but this time, with definite malicious intent. In the Netherlands, a worm nicknamed "Duh"[H] started spreading to jailbroken iPhones with default root passwords in attempt to steal banking information. The worm would setup a botnet where a bot residing on the phone would wait for SMSes from ING bank and mobile transaction authentication numbers (mTANs), a six-digit number sent to Web banking users that they must enter in a short window of time to complete a transaction. An out-of-band authentication system to verify the owner is accessing the account. The system fails since the SMS is received on what amounts to the same computer that is accessing the Web banking portal, thus negating the out-of-band part and the security it implies. The bot intercepts the mTAN SMS and sends it to a remote system where the attacker has a small window of opportunity, logging in ahead of the legitimate user and gaining access to their funds. Though not widely spread, any successful attacks would have likely yielded enough funds to make the effort worthwhile to the attacker and likely would inspire similar attacks in the future.

FUTURE ATTACKS

By no means is the future of attacks limited to RickRolling and bank fraud. The potential for these devices as vectors for other attacks is huge. The one constant within security is that threats are continuous and often will surprise everyone. Smartphones and hybrid devices are so very versatile that people are constantly coming up with new ways of applying these advantages in some surprising ways.

Twitter founder Jack Dorsey started a new venture called Square (www.squareup .com). The company hopes to become a major player in mobile credit card payment systems. The company is developing a small credit card reader that plugs into the audio in jack of a smartphone like an iPhone and allows anyone to accept credit cards for purchases without the cost and burden of traditional credit card authorization systems.

Motorola is also entering the field with their mobile payment module (www.motorola .com/Business/US-EN/Business+Product+and+Services/Accessories/Mobile+ Computer+Accessories/Snap-ons/Mobile+Payment+Device/Mobile_Payment_Device_ US-EN), which clips onto the bottom of a smartphone running Windows mobile to accept and process payments anywhere. Using existing commodity devices, lowers costs are associated with the equipment.

[H]www.sophos.com/pressoffice/news/articles/2009/11/iphone-worm-duh.html

While these ventures are still in early stages, this does make for some interesting thoughts on hybrid device security. Since these devices now accept credit cards, they are a more interesting and valuable target for an attacker. No longer would an attacker just get away with contact lists and e-mail – he or she may be able to retrieve credit card information in bulk from these handheld devices.

The technology is still evolving but Square and others are leveraging, in part, the ubiquity of the iPhone and its interface. Credit cards are entered by hand on the phone or through a reader connected to the phone. The transaction is routed over 3G or Wi-Fi to be authorized in real time. While this may seem convenient since you can now process credit cards from anywhere, it also presents an interesting angle for an attacker. iPhones are very easy to coax onto Wi-Fi networks with tools like Karma, and given their preference to automatically connect to known networks, this would override the 3G connection. It would take a very small amount of network routing over an alternate backhaul, and all the traffic would be routed through the attacker. Combine that with existing weaknesses in the secure sockets layer and how users deal with error messages, and the attacker has a decent chance of being able to record all transactions. The closed nature of the iPhone causes issues in this scenario, as well, since the application vendor cannot "peek under the hood" and make adjustments to the device to secure it further.

This is still a theoretical situation as no major attacks have been noted (or at least noticed) via this or other methods. That said, there is nothing preventing a "perfect storm" of new vulnerabilities, slow patching cycles associated with cell phones, and opportunities that would not make for a very large and profitable attack.

Offensive Uses of Hybrid Devices

There is also little assurance for those paying via such mobile payment systems that the devices have not been altered in some way. Perhaps an unscrupulous employee jailbroke the device and altered the device to send a second copy of the transaction to his or her personal e-mail. In the same way, the Ikee and Duh' worms used jailbreak access to change system files without the users' permission; the same access could be used by the user to alter the system and compromise others without their knowledge.

Another angle not often considered is using hybrid devices as a tool for attack, rather than as the target. Walking into a target location with a laptop and a bunch of other gear may be fairly suspicious and get a fair bit of unwanted attention for the attacker, but walking in with an iPhone and tapping at its screen is something that is highly unlikely to be considered suspicious.

At Defcon 16 in 2008,[1] David Maynor and Robert Graham of Errata security gave a presentation called *Bringing Sexy Back: Breaking in with Style* in which they described a number of methods they had used to creatively break into companies during penetration tests.

As part of the presentation, the duo discussed using an iPhone as a sort of Trojan horse to gain access to a facility and assess its internal wireless infrastructure. Inspired

---

[1]https://media.defcon.org/dc-16/video/Defcon16-Maynor-Graham-Bringing_Sexy_Back.m4v

by a *Saturday Night Live* sketch, the pair got the idea of overnight shipping a modified iPhone to the target facility in order to gain access past the perimeter. In most cases, mail and package deliveries just pass right on into the facility and into the mail room, right into range of any wireless infrastructure that is present at the facility.

They first obtained an early generation iPhone (due to lower power requirements) and the original box. The original iPhone battery would only power the device for a day or so, which means it would already be dead upon arrival. To combat this, they installed a battery pack in the bottom of the box and used that to power the iPhone for up to five days, more than enough time to do what was needed. With a little modification of the box, the battery, charge cord, and iPhone all fit snugly in a nonthreatening package.

The iPhone was jailbroken and additional software installed well beyond anything Apple intended to ever have installed. The iPhone, being a UNIX-based system, made porting of several popular utilities easy to do over other platforms (like Windows Mobile or Symbian). Tcpdump (packet sniffer), Metasploit (exploit framework), FERRET (information leakage assessment), and other tools were installed on the device. After disabling all the auto power down features of the phone, they setup the system to call home. Since most carriers block incoming connections to devices on cellular data services, the phone was made once an hour to connect out over the cellular data connection to a remote system and establish a shell for the penetration testers to use.

Once everything was set up and charged, they overnight-shipped the box and phone to the test facility to the attention of a nonexistent employee name. Presumably, the shipment tracking number told them when the package arrived at the destination and when they could begin testing. Addressing the package to a nonexistent employee was a way that the package would stay in the mail room for a period of time since it was undeliverable and (hopefully) be returned to Errata Security after a few days to be recharged and used again.

This attack proved to be very effective and continues to be developed. In the hands of a determined attacker, it could be devastating as it also includes a level of anonymity that would be highly desirable.

NOTE

Since the iPhone was jailbroken, this removed the digital locks put in place to prevent using the phone on other networks. An attacker could then use an anonymous prepaid SIM card from some far away land that would be very difficult to track down. As well, if the package was opened out of suspicion, there is nothing really indicating who sent it (other than a return address which may or may not exist), and the attacker is only out a few hundred dollars and is safely far away from the facility.

Once inside the facility, the Errata penetration testers had everything they needed. They had remote access to the device that had a wireless interface they could listen to traffic on, use to connect to the network, and a UNIX platform with tools like Metasploit installed to conduct further tests and attacks on, all while this small box sat on a desk or shelf in the mail room.

This method of attack was fairly ingenious, as a common defense used by some is to have a heavy outer layer of security (guards with guns), assuming that this will prevent an attacker getting close enough to take advantage of poor security further in.

The following year, at Defcon 17[J] in 2009, a presentation by Thomas Wilhelm expanded on the capabilities of Errata Security by exploring the possibilities of hacking with the iPod Touch (similar to the iPhone in capabilities, but without the telephone functions). A surprisingly large amount of tools have been ported and are simple to install on a jailbroken devices. Some others require manual building installation, but all the tools to do so are easily available. He noted that an iPod/iPhone is more socially acceptable to bring into a "secure" area than a laptop or other obvious computer.

Anonymity

The potential with these devices is not just limited to them being a target of attack or a tool of attack, but as intermediaries to do the dirty work through or for the attacker, unknown to the user.

It's not hard to imagine a day when a remote vulnerability on an iPhone will allow an attacker to gain root access to a device over local Wi-Fi connections. Once there, all the tools necessary to download and install further malware already exist. An attacker could set a scheduled job that can run who-knows-what at a later time.

It is not also unreasonable that an attacker could use the Wi-Fi connection as a gateway to access the Internet through the cellular data connection. Previously, an attacker trying to hide his/her tracks would need to route their connection through compromised computers around the world or an open Wi-Fi access point. The problem with these is that connections can be logged and some sort of trail left. In the case of a mobile phone, there is no facility to login that kind of detail what connections are routed through it leaving the attacker to just have to visit a coffee shop in order to do their evil deeds.

iPHONE SUMMARY

As a growing part of the market, iPhones are going to be under greater scrutiny by attackers for vulnerabilities, particularly as more and more people put more and more personal parts of their lives onto these devices.

For now, attacks have been limited to jailbroken devices left with default passwords. However, we should learn lessons from these experiences because, at some future point, a vulnerability may be found in a default application, or even a third-party application sanctioned by Apple, that may provide an attacker with a path to the rest of the system.

[J]https://media.defcon.org/dc-17/video/DEFCON%2017%20Hacking%20Conference%20Presentation%20 By%20Thomas%20Wilhelm%20-%20Hacking%20with%20the%20iPod%20Touch%20-%20Video%20 and%20Slides.m4v

The closed nature of the iPhone is a double-edged sword for security. Being that it is closed and applications vetted before being offered through the App Store, Apple can limit the amount of access applications have and hopefully limit exposure to malware. This comes at a cost to the user, though. Because it is closed, the user is limited in what sort of proactive steps can be taken to fend off attacks. Currently, there are no firewall applications available for the iPhone to filter what type of traffic is allowed (incoming or outgoing). Virtual private network (VPN) functionality is built in, but it is limited in support for different protocols. If you have an existing VPN infrastructure that does not support any of the included protocols (Layer 2 Tunneling Protocol, Point-to-Point Tunneling Protocol, or Cisco IPSec), you are out of luck. There is also no available option for an antivirus scan of the device as that would almost be an admission by Apple that their devices were vulnerable in their factory (nonjailbroken) state (something unlikely to happen).

Android Security

While iPhones enjoy a growing percentage of the mobile phone market and a certain amount of sex appeal, phones based on Google's Android OS are growing in interest and market share. Unveiled in late 2007, the OS, which is based on a Linux kernel (as opposed to iPhones FreeBSD-based Darwin kernel), has been built with a different security model than that of the iPhone. The open nature of the OS and its attention to security is a significant counterpoint to Apple's closed model.

Android's security model is one of openness. This seems to be counterintuitive in that security and openness are concepts usually not considered compatible. Android developers' desire was to allow users and application developers to have a greater degree of control over the device, and to put in place a model where even if a malicious application was installed or a remote bug found, the implications would not be as dire.

On Android, much of the security comes from a unique model for software permissions. Each program runs as its own User Identifier[K] (UID or simply user), whereas on an iPhone, all programs run under a single user (mobile). This acts to "sand box" and limit interprogram communication to specific functions. At the time of install, the program can request specific permissions to interact with other programs. The user is told of these permissions and has the option of allowing or disallowing of these permissions. In this way, the user can decide that, yes, this new contact manager should be able to access my phone book and make calls, but not the Wi-Fi network or the GPS. The same goes for programs that have no business accessing core phone functions. Why does this falling block game need to know my location and access my phone book? In this model as well, a potentially malicious program can only interact with other programs and data in very specific ways, and as such, limits the amount of damage that can be done should they go rogue. Applications can also generate their own permission types to allow other programs access.

[K]http://developer.android.com/guide/topics/security/security.html

In the Android model, the user is the gatekeeper who can make decisions about security. In the iPhone model, Apple is the gatekeeper. All apps have to be submitted to Apple and pass through a screening process before they are added to the App Store. Apple checks the source for obvious things, such as a game accessing the microphone or call functions and other things that it should not be doing, but given that Apple receives about 10,000 applications and application update submissions a week[L] and has only 40 full-time reviewers,[M] the possibility of the process being overworked and missing something or for overzealousness and unnecessary denials happening[N] is real and has occurred. Android leaves it up to the user to decide. Offensive or unstable applications will simply be uninstalled by the user and even then, due to permissions, unstable or malicious applications will be unlikely to affect the core of the system and other applications.

On the iPhone and other platforms, a program with malicious intent or one compromised in some way is capable of accessing data and resources from any program running under the same user. Once the program breaks out, there is no restriction other than Apple's guidelines in the App Store agreement. This means that a compromised mail client can, in theory, access contact lists and issue SMSes, which in turn would cost the user money. The only thing stopping this is that Apple will not allow such an application to be approved and end up in the App Store.

Common Threats

While Android and iPhones are fairly different in the model they use for application security, they share some fundamental similarities that make smartphones and hybrid devices as a whole market a much more interesting target for attackers than any one device or brand alone.

A common argument against extra security for smartphones is that the amount of any single model or platform of phones in comparison to other targets is relatively small. An install base of a few million Android-based phones is not a very large target base when compared to the massive number of Windows-based internet connected PCs. A vulnerability exploited on Windows PCs has a much higher chance of mass success than just a few million smartphones, not all of which are on or connected.

However, it is worth remembering that these phones are computers in their own right and as a whole represent a very large number of devices, even outselling laptops in the foreseeable future.[O] Put together, smartphones are a sizeable install base and as different as they seem, beneath the hood they share more than one would think.

To save costs, increase developer adoption, and other factors, modern smartphones are now leveraging existing UNIX platforms instead of writing entirely new

---

[L]www.businessweek.com/technology/content/nov2009/tc20091120_354597.htm
[M]www.apple.com/hotnews/apple-answers-fcc-questions/
[N]www.alkalimedia.com/2009/04/21/
[O]http://ce.tekrati.com/research/9697/

OSes. Much of this includes open source tools and libraries. Despite their outward appearances, the iPhone, Android OS, and even the Palm Incs WebOS, which powers the Palm Pre, share a number of common programming underpinnings, which mean that there is potential for all devices to be vulnerable.

One such underpinning is WebKit,[P] the engine used by iPhone for the Safari browser and e-mail client; Android uses it in its browser and WebOS uses it heavily throughout the OS to render applications. This common engine is a fundamental part of all three of these new generation smartphones.

This creates an interesting situation because, since the engine is used in the browsers in all three, there is the potential for an exploit to work on all three platforms, greatly increasing the target base and removing much of the need to tailor the attack for specific platforms. Usages common between the platforms as well make attacks easier. Tools like Google's Gmail and other Web-based applications are very popular with mobile users and are also under considerable attack.

An exploit in WebKit, like a cross-site scripting bug, could generate an iFrame and be used to upload the contents of these Google applications to the attackers server of choice on any platform. Since even with Android's permission structure the browser needs access to the network, this sort of action would be allowed. Combined with large crowds like at a mall or conference and client attacks like using Airpwn to inject malicious scripts into traffic, an attacker can attack users across any platform, not just one.

There are likely more commonalities between platforms than just WebKit, and attackers are notoriously creative in leveraging vulnerabilities. Should such vulnerability be found, they would certainly love to expand their potential attack base.

SUMMARY

While the future rarely arrives exactly as predicted, some elements of it do make it to market. The advance of the mobile phone from simple communications device to hybrid mobile computing platform has been amazing to see in a fairly short period of time. The resulting drop in cost over the years, allowing everyone to own one, has changed our day-to-day behavior in ways most people never expected.

The threats to and from these devices are a new challenge for society to adapt to. While manufacturers can build in security mechanisms to attempt to think ahead of attackers and cut them off at the pass, there are still a great many ways that things can go wrong. Technological development will not slow down in order for society to catch up. Consumers of these devices need to take more forward steps to understand the implications of the devices they use and how they can affect their daily lives.

[P]http://webkit.org/

Index

If you've enjoyed reading about these attacks you will love *Seven Deadliest Network Attacks*, another book from our Seven Deadliest Attacks Series.

Password Replay

It seems that in 2003 (although the exact year differs in different accounts) hackers aimed an antenna at a Marshalls clothing store near St. Paul, Minnesota, in order to capture data from the store's wireless network. By capturing network traffic and analyzing it, the hackers were able to obtain the wireless network password. Once they were in, they were able to "sniff" for other passwords on the network, eventually obtaining access to the databases of parent company TJX in Framingham, Massachusetts. Ultimately, the hackers downloaded around 94 million credit card numbers.[A] They also got personal information, including Social Security numbers, for about 451,000 customers. TJX personnel discovered the intrusion in December 2006, and found files left behind by the hackers. Perhaps ironically, TJX cannot break the encryption the hackers used on the files. The cost for TJX? Aside from significant embarrassment the total cost could top $1 billion over 5 years to pay for security consultants, lawyers, and of course marketing to help reassure customers. Of course, this figure does *not* include potential liabilities from lawsuits.[B] Who will sue? Affected banks, for one, are alleging negligence in TJX security practices. There's also the Federal Trade Commission (FTC). Sometimes life is tough.

Much of this intrusion was accomplished using the technique of *password replay*. The security used on the wireless network, wired equivalent privacy (WEP), was

---

[A]For the 94 million figure, see Ross Kerber, "Court Filing In TJX Breach Doubles Toll," *Boston Globe*, October 24, 2007.
[B]There are many articles on the TJX break-in. In particular, see "How Credit-Card Data Went Out Wireless Door," *The Wall Street Journal*, May 4, 2007.

cracked using a variant of password replay, and then passwords were captured from the unencrypted network traffic. This allowed the intruders to make their way along the TJX networks and copy out the contents of databases. Even better, they were able to install software at points in the network to "sniff" the unencrypted credit card data traveling the network.

Of course this was all years ago. On January 20, 2009, Heartland Payment Systems (HPS) announced a bigger breach. HPS employees discovered programs on their network to capture traffic. In this case, the thieves compromised an estimated 100 million credit cards. This information isn't just the credit card numbers, but includes the full data in the magnetic stripe of the card, allowing the thieves easily to create duplicate cards.[C]

The good news? It looks like the mastermind of both operations has been caught. Albert Gonzalez has pleaded guilty.[D] He leaves behind a record of computer break-ins and $1.1 million in cash discovered wrapped in plastic and buried in a drum in his parents' backyard.[E] Again, sometimes life is tough.

HOW PASSWORD REPLAY WORKS

It's 1995, and you're sitting in your office arguing in an online newsgroup about the new show *Star Trek: Voyager*. Your "friend" Rob stops by and you chat for a while. Before he leaves Rob asks if he can use your terminal to check his e-mail. "Sure," you tell him. Rob logs in, checks his e-mail, and the logs out. Or so it seems.

After Rob leaves you decide to log in and check out this new online bookstore you've heard about. You log in, and you are told your password is incorrect. You've mistyped it before, so you type it again, and it is accepted. After visiting the bookstore's Web site you decide "Hmph. You can't actually flip through the books, and the Web is too insecure. This will never catch on."

Not only are you wrong about the bookstore but also Rob now has your password. You've just been the victim of a fake login that allowed Rob to capture your password. It's shockingly easy to do this in most cases. For example, if you are using Ubuntu 9.10 and use console login instead of the graphical login, then the script shown in Figure 7.1 will emulate the log-in prompt, capture the next username and password, report that the password is in error, and then terminate itself. The end result looks just as it should. People happily accept that they have mistyped their password once, and so no suspicion is aroused.

You'd log in on the first text console (named "tty1") and run this script. The effect is as if you had logged out; but of course, you haven't. The script prints an official-looking

[C]Rachael King, "Lessons From the Data Breach at Heartland," *BusinessWeek*, July 6, 2009.
[D]"Man Accused of Stealing Stores' Data Pleads Guilty," (REUTERS), *The New York Times*, August 28, 2009.
[E]Scott Hiaasen, Rob Barry, Nirvi Shah, and Michael Sallah, "From Snitch to Cyberthief of the Century," *The Miami Herald*, August 22, 2009.

```
#!/bin/bash
clear
echo ""
while [ -z $login ] ; do
  echo "Ubuntu 9.10 ubuntu tty1"
  echo ""
  read -p "ubuntu login: " login
done
stty -echo
read -p "Password: " password
stty echo
echo ""
echo "Login incorrect"
echo "$login:$password" > capture.log
kill -9 $PPID $$ >/dev/null
```

FIGURE 7.1

A Very Simple Fake Login

log-in prompt and captures the username and password (character echo is turned off for the password so that things appear correct). The script then claims you have entered the wrong credentials, saves what you *did* enter, and kills itself and your log-in shell. Presto! You're logged out and the data is captured. Your victim now gets a *real* log-in prompt and can log in.

University computer labs used to be (and some still are) full of UNIX workstations with console logins, and this was a not-too-uncommon occurrence. There are advantages to having a login that is not your own, especially if you are planning to do something dastardly. System administrators would sometimes watch

---

NOTE

You don't use a console UNIX login, so you're golden, right? Well, no. Graphical user interface logins can be spoofed as well, and it is both easy and effective. Just write a program that uses the whole screen to display the appropriate login, and then capture the data entered. On some systems, you can use this information to actually log the person in, so when they log out your "fake" login is ready to accept the next victim.

Microsoft has addressed this in more recent versions of Windows by requiring the user to press **Ctrl + Alt + Del** to initiate a login. This is a good idea, as this key combination cannot be trapped by the fake login. Or can it?

There are actually several approaches to trapping the **Ctrl + Alt + Del** combination that work just fine. After all, this is often done for "kiosk" installations of Windows, where a single application has to be running all the time, no matter what. These require that the computer user has administrative access to the operating system, so he or she can modify the keyboard driver, modify the registry, or (for versions earlier to Vista) manipulate the graphical identification and authentication (GINA) library. There's an open source implementation of GINA for you to start your hacking: pGINA.[F]

[F]See http://sourceforge.net/projects/pgina/.

for unoccupied machines with someone still logged in; it could be someone who forgot to log out, or something more nefarious. The "solution" to these watchful eyes is trivial social engineering: wait until the lab is full, start the program, and then kindly hand off the terminal to the next person who is waiting. The script logs you out and then the other person logs in; everything is just as it should be. Well, mostly.

Although this isn't an example of a *network* attack, it illustrates an important point about security. Passwords are typically *static* and *trusted*. That is, they don't change (at least, not very often), and they often serve as a single point of authentication that you are who you say you are. They are similar to physical keys or proximity cards ("prox cards" or "key cards") in this regard. If you have the key or the proximity card (or a duplicate), you have access.

The important distinction between passwords and physical keys or proximity cards is that the latter is intended to prevent physical trespass; you have to go to the location, and there may be guards who do not recognize your face, cameras to take your picture, and suspicious employees. You probably have to get dressed. Life is sometimes tough. Passwords prevent *virtual* trespass, and you may be able to accomplish this from the (relative) comfort of your parent's basement while wearing your bathrobe. Another important distinction is that the information you really, really want is likely stored on computers, anyway. Why get off the couch if you don't have to?

WARNING

Fake logins are *very* common on the Internet.[G] *Phishing* is a very common attack that combines social engineering with password capture to steal log-in credentials, credit card numbers, or other personal information. Most phishing e-mails direct you to a Web site purporting to be your bank, credit card company, or employer. When you enter your credentials or other information on the site, however, it is captured and saved. You may then be directed to the real site.

The best way to detect this sort of attempt is to carefully examine the Web address (URL) in the message and see if it is really the site you expect. The best way to avoid phishing scams is to simply *never click on a link in an e-mail* unless it is from someone you trust or in a communication you are expecting (such as a registration confirmation e-mail). Even then you should examine the link; it is possible that a "man in the middle" (see Chapter 6, "Man-in-the-Middle") could rewrite the link to point to their site. Legitimate organizations should never send you e-mail with instruction to click a link and enter personally identifiable information. If you need to log in to your bank, enter your bank's address or use a bookmark, and then log in.

[G]A fake Google mail (Gmail) login is described at http://www.trap17.com/index.php/fake-gmail-interface_t57079.html (retrieved on December 1st, 2009). The interesting thing about this fake login is just how similar it is to the UNIX command-line attack described previously in this section. After the fake login captures your credentials, it forwards you to the *real* Gmail site. Note that the fake login has long since been taken down.

Simple Password Sniffing

One way to capture passwords on a network is to capture and decode network packets. As discussed in Chapter 5, "Spanning Tree Attacks," Wireshark,[H] tcpdump, and libpcap[I] can be used to capture packets. If you have access to a node through which the packets will travel, and the passwords are unencrypted, you can capture them very easily. Note that this is a form of the "man in the middle" attack detailed in Chapter 6, "Man-in-the-Middle."

This method of password capture assumes two things are happening.

1. Static password authentication is required.
2. Passwords are being sent as "clear text," that is, unencrypted.

Surely nobody would do this, you might say. Well, they do. For example, this might be done if the communication channel is itself encrypted, as with a virtual private network (VPN). This works great to secure traffic between your machine and a remote network; traffic passing through intermediate nodes is encrypted. It does not work so well if someone has access to the remote network or, in some cases, to your machine. Once someone breaks into the remote network by some means (see the

---

NOTE

What's your password policy? There are approximately 95 characters that are typically usable in passwords. Letting a user choose *any* password gives a grand total of more than six *quadrillion* possible passwords of length up to eight. That's actually a lot, and includes such gems as "password" and "0),,!*fZ".

We believe (with good reason) that the password "0),,!*fZ" is harder to crack than "password" because many password-cracking tools use dictionary-based attacks, and so we establish rules to promote stronger passwords. We might require at least one digit, at least one lower- and one upper-case letter, and at least one nonalphanumeric character. This limits our choice on four characters, and restricts the realm of possible eight-character or shorter passwords to roughly 18 *trillion*. That's much, much fewer, and is pushing us into the realm of effective brute-force attacks. In fact, it becomes possible to precompute tables that can instantly crack any such password.

This is a tension between usability and security. Is the password "0),,!*fZ" better than, say, "hgqpmngd"? Well, keep in mind that some password-cracking tools such as Cain & Abel,[J] described in the later section "Password Replay," explicitly include code to run a brute-force attack on lower-case-only passwords. It is, of course, possible to go completely wrong with your policy. The more restrictive you make your policy, the smaller the search space for an attacker.

Requiring your users to change their passwords periodically and to avoid reuse of old passwords can be a good policy. They are less likely to change their passwords everywhere else, so it is highly unlikely that their account on your enterprise resource planning (ERP) system and their account on startrek.com share the same password. That's a good thing.

[H]See http://www.wireshark.org/. Note that this tool was previously named Ethereal.
[I]See http://www.tcpdump.org/ for tcpdump (a command line tool) and libpcap (a library).
[J]See http://www.oxid.it/cain.html. Users of non-Windows operating systems should have a look at dsniff: http://www.monkey.org/~dugsong/dsniff/.

rest of this book) then they can sniff packets and look for clear text passwords. This gives them access to other accounts, possibly to other data, and perhaps even to your accounts on *other* networks. Do you use different passwords for all your accounts? Do you ever reuse old passwords? If somebody has a list of the passwords you used last year they can use these to start guessing your current passwords.

Running a tool like Wireshark on a network node and watching for Post Office Protocol (POP) or Simple Mail Transfer Protocol (SMTP) traffic[K] can often reveal interesting data. Of course, combing through all the captured traffic can be burdensome. Can't someone automate that task for us? Of course they can. EffeTech makes a commercial password sniffer called Ace Password Sniffer,[L] and NirSoft SniffPass[M] is a freeware utility that captures packets and watches for passwords in commonly used protocols. These are just two; a quick Web search for "password sniffer" can turn up many, many more. These are typically advertised as a way to recover lost passwords or monitor children on the Internet, but as with most tools, they have other uses.

WARNING

Suppose you're running Wireshark. You may be trying to capture packets in an attempt to break into a machine, but more likely you are a network administrator trying to monitor your network. Wireshark has to read and respond to essentially arbitrary data, and typically does so while running as the "root" administrative user (so it can capture all traffic). Security vulnerabilities in Wireshark, or any other such program, can actually lead to an attacker gaining control of the machine where Wireshark is running! Keep your software up-to-date and pay attention to suggestions for securely running software like Wireshark that needs special privileges.

All these depend on your ability to "sniff" the traffic. That is, the traffic has to pass through (or be available to) your machine. There are a variety of ways to get the traffic. The most straightforward is just to install the password sniffer (or a traffic-capture program like tcpdump) on a gateway or proxy server. You can then watch all the traffic passing through the machine on its way to and from the Internet. It may even be possible to modify a network's topology (see Chapter 5, "Spanning Tree Attacks") so that your machine receives all traffic. You can only sift through the traffic you actually see, after all.

Of course, sometimes people decide to broadcast all their packets to the world. The mechanism to accomplish this heightened level of insecurity is called "wireless networking," and you can often find unencrypted wireless networks in coffee and sandwich shops. Although you might not be able to break into some well-secured corporate network, you might discover that the vice-president has an unsecured (or poorly secured) wireless network at home, and he tends to use the same passwords

---

[K]POP is a very common protocol for mail delivery (getting the mail from the server to you), while SMTP is a very common protocol for sending mail (getting the mail from you to the server).
[L]See http://www.effetech.com/aps/.
[M]See http://www.nirsoft.net/utils/password_sniffer.html.

for everything, from his corporate e-mail to protecting his VPN certificates. Using his e-mail account, you might even compose angry e-mails "from him" to the corporate IT department. Who knows what you might accomplish this way?

Password Replay

You've got the access to a switch that is carrying your "friend" Rob's Internet traffic, and you are happily collecting packets. You see that he occasionally connects to a remote site and you are *sure* he is busy extolling the virtues of Kirk as captain of the *Enterprise* and running down your recent postings about Picard. You have to know for sure, but you can't capture any clear-text passwords. The system he is using encrypts his credentials when he sends them, so you can't just grab them from the network traffic. What can you do?

You may be able to *replay* the packets. Replay attacks work by first recording an authentication session, and then playing that session back at a later time. Using this strategy you may be able to observe Rob's authentication session, and then replay the recorded packets at a later time to gain access as if you were Rob.

Recording and playing back packets sounds like something that requires programming. Luckily there are ready-made tools such as tcpreplay[N] to automate most of this process for you. Actually, "tcpreplay" is a suite of tools for classifying, editing, and replaying network traffic. These tools work from a "pcap" file containing captured traffic, created with another tool like tcpdump. The tools are quite sophisticated.

"Sophisticated" might sound like another word for "hard to use." All you want to do is get Rob's password. Do you really have to capture traffic, sift through it to find the kind of packets you want, extract those (you can use tcpdump to refilter an existing pcap file), classify the packets so you get the client (Rob's) traffic, edit the traffic if necessary to modify the IP addresses, and then replay the traffic? Whew! Stealing— I mean, "recovering" passwords from network traffic must be a common activity. Isn't there an easier way?

Of course there is. One of the best tools around for this sort of work is a freeware tool called Cain & Abel. This is one of those cases where Windows users have an exceptionally powerful tool that really isn't available for other operating systems. After you've downloaded and installed Cain & Abel (but see the TIP box first!) you are ready to begin capturing passwords, conversations, and other network traffic.

Replay is another exploitation of a static (or at least predictable) authentication system. If the challenge and response depend on a sequence number that is not predictable by the eavesdropper, then password replay will probably fail. In fact, including a cryptographic sequence number is the most common means to prevent password replay attacks. Given this, you might think that practical password replay is a nonstarter—but you'd be wrong. Many systems are susceptible to replay attacks. You may have some of these systems in your infrastructure right now.

[N]See http://tcpreplay.synfin.net/trac/.

> **TIP**
>
> Cain & Abel is a well-known password sniffer. In fact, it is so well known – and so effective – that antivirus and antimalware vendors detect it. You may not be able to install it if you have an active antivirus program on your machine running in an "auto protect" mode. Antivirus scans might discover the program and damage or remove it by trying to "quarantine" parts of it. They may either prevent the "Abel" service from starting, or detect and kill it. How rude! All you want to do is capture passwords. What's so wrong about that? The lesson is that you should disable your antivirus software before you download, install, and run Cain & Abel.
>
> The Cain & Abel software expects to be able to intercept packets. An active firewall's rules can interfere with some aspects of the program, so you might want to also disable the firewall before you run the program. Disabling the antivirus and the firewall on a machine can be dangerous—but Rob should have thought about that before he started rambling on about *Star Trek*, something he clearly knows nothing about. It's his own fault, really, that you had to install a password sniffer on his machine.
>
> Installation of Cain & Abel requires reading the instructions; you have to install and start the Abel service separately from the Cain front end. There is also a method to remotely install the Abel service and start it on another machine, and then connect to it with the Cain interface.
>
> If you install Cain & Abel on a machine used by others, they may detect it. By default Cain & Abel installs the registry key HKEY_CURRENT_USER\Software\Cain. In short, the program is fairly easy to detect, and isn't really intended to be hidden. Perhaps after you are done using it, you may consider running the uninstaller that comes with it. It is, after all, courteous to clean up after yourself.

What about encrypted traffic and passwords, or protocols using sequence numbers? It might be surprising to know that these can also fall prey to replay attacks for several reasons:

- The protocol might be cryptographically weak.
- The protocol might have a fundamental weakness that exposes credentials.
- It may be possible to use a man-in-the-middle attack to overcome the encryption.

An example of a cryptographically weak protocol is WEP,[O] a protocol used to secure wireless networks. Sadly, the protocol is constructed in such a manner that it is possible to quickly break the encryption by capturing special packets called *initialization vectors* (IV). This attack was used in the TJX break-in described at the start of this chapter.[P] Several tools exist that can be used to sniff packets, collect IVs, and then crack the wireless password. Because it can take a while to collect enough packets, these tools commonly support packet *injection*, where the attacking machine generates traffic to cause the wireless hub to generate new IVs. The Aircrack-ng tools[Q] provides wireless network cracking under Windows and Linux for WEP, as well as

---

[O]See IEEE standard 802.11-1997.
[P]See Larry Greenemeier, "T.J. Maxx Data Theft Likely Due to Wireless 'Wardriving,'" EE Times, May 9th, 2007. http://www.eetimes.com/.
[Q]See http://www.aircrack-ng.org/.

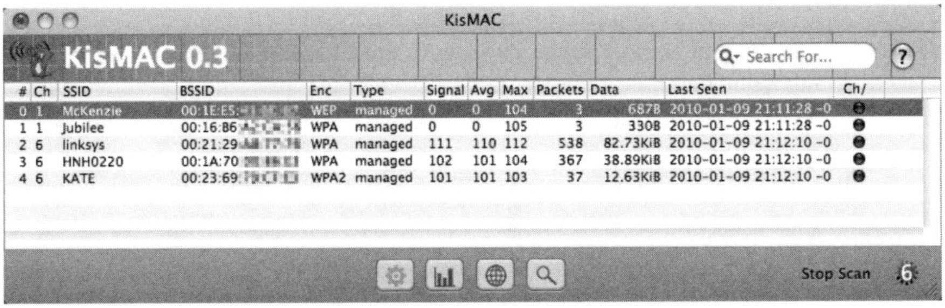

FIGURE 7.2

KisMAC

the more modern wi-fi protected access (WPA). Mac users have the KisMAC tool.[R] Figure 7.2 shows KisMAC running and collecting packets for several networks. Once enough packets have been captured, the cleartext WEP password can be cracked.

Server Message Block (SMB) is a protocol for network communication between network nodes and to shared devices such as printers. SMB is the application-layer network protocol of the Microsoft Windows network and is used throughout the Windows world. NT Lan Manager (NTLM) is an authentication protocol used with SMB in Windows versions earlier to Vista. (It is still present in Vista, but deprecated. Kerberos[S] is the new authentication system.) There is a chance that, at the time of writing, your infrastructure may still be using a version of this authentication protocol with a fundamental weakness: it honors remote requests for authentication. Suppose you receive an e-mail from your "friend" Rob inviting you to join a Kirk vs. Picard discussion, and providing you a link. You immediately click on the link to give everyone the benefit of your opinion. When you click the link, you are connected to

---

EPIC FAIL

SMB has been known to be "broken" since 2001, but changing a network protocol is a nontrivial matter. Lots of devices and network-based applications depend on the protocol implementation, and changing them all at once isn't really an option. You don't want your e-mail to quit working, do you? Microsoft kept working on a way to fix the problem, eventually releasing patch MS08-068[T] in November 2008. So you only had 7 years to exploit this particular vulnerability. Of course, it took until July 2007 to implement the exploit in the Metasploit 3 framework.[U] At the time of writing, there are other outstanding security issues that Microsoft is currently working on. As you read this, that is probably still true.

[R]See http://kismac-ng.org/.

[S]Kerberos is a very common authentication protocol under Windows, Linux, and UNIX, including OS X. See http://web.mit.edu/Kerberos/.

[T]See http://www.microsoft.com/technet/security/Bulletin/MS08-068.mspx.

[U]See http://www.metasploit.com/. Metasploit was created in 2003, so it only took four years to get around to writing that particular exploit.

a server that requests that you (the client) authenticate yourself using NTLM. NTLM responds by happily sending your credentials to the server, which stores them. Later Rob logs into your machine with the stolen credentials and changes your desktop wallpaper to an image of Picard and even goes so far as to delete your fan script. People can be so mean.

Finally, it is possible to use a variant of the man-in-the-middle attack to capture and replay passwords even in the presence of encryption. Precisely how to do that is the subject of the following section.

Address Resolution Protocol Poison Routing

In Figure 7.3, Alice wants to connect to the server out on the Internet. Her traffic flows through the gateway, to which Eve is also connected. We see the Internet Protocol (IP) and media access control (MAC) addresses for the machines. Chapter 5, "Spanning Tree Attacks," discusses exploiting the Spanning Tree Protocol, a "layer 2" protocol. Another layer 2 protocol is the Address Resolution Protocol (ARP). ARP is used to map between IP addresses and MAC addresses. Although a network card might be assigned any of several IP addresses over the course of its life, it (typically) has a single permanent MAC address.[V]

Alice wants to communicate with the server, but it is on a different network. She therefore sends her traffic to the gateway with IP 10.1.2.1. Her machine's

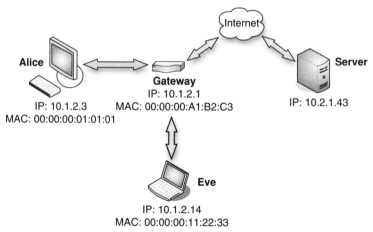

FIGURE 7.3

Preparing to Eavesdrop

[V] Of course, "permanent" here takes on its computer science meaning of "not really permanent." That is, it is possible to assign the card a different MAC address. This is called MAC "spoofing," and it has a variety of benevolent and malevolent uses.

ARP tables indicate that this IP address belongs to the machine with MAC address 00:00:00:A1:B2:C3. The gateway then takes care of forwarding the traffic on to the Internet. Likewise, packets arriving at the gateway for Alice's machine are mapped to MAC address 00:00:00:01:01:01.

Eve wants to listen to the traffic between Alice and the server (or between Alice and the whole Internet, for that matter). To accomplish this, Eve uses ARP "poisoning," sometimes (confusingly) referred to by the acronym APR for ARP poison routing. Eve sends out an ARP update to Alice's machine at 00:00:00:01:01:01 pointing the IP address 10.1.2.1 to Eve's MAC address 00:00:00:11:22:33, and Alice's machine dutifully stores this in its cache. Eve then sends out an ARP update to the gateway machine at 00:00:00:A1:B2:C3 pointing the IP address 10.1.2.3 to Eve's MAC address 00:00:00:11:22:33. Eve's machine can still route traffic to the gateway and to Alice's machine using the correct MAC addresses. Now Alice's machine thinks Eve's machine is the router, and the router thinks Eve's machine is Alice's.

Alice wants to log into the server, so she sends a request to the server. The server is on a different network, so her machine determines that it needs to be sent to the gateway at IP 10.1.2.1. Alice's machine looks in the ARP cache and finds MAC address 00:00:00:11:22:33, and sends the packets to that MAC address. In this case, the gateway is connected to Eve's machine, but the packets are labeled for 00:00:00:11:22:33, so the gateway sends them on to Eve's machine. Eve can now modify the packets however she wants and then send them to the gateway at MAC address 00:00:00:A1:B2:C3. The gateway is the destination for these packets, so it examines them, and determines whether they should be sent on to the Internet.

Next the server replies to Alice. Packets arrive at the gateway destined for IP address 10.1.2.3. The gateway looks in its ARP cache and determines that this IP address belongs to the machine with MAC address 00:00:00:11:22:33. It then sends the packets on to Eve's machine. Eve is now free to modify the packets however she wishes, and then she sends the modified packets on to MAC address 00:00:00:01:01:01. The gateway receives these packets that are not for it, and dutifully forwards them on to Alice's machine. Eve has successfully become the (wo)man in the middle.

Now, Alice wants to establish a secure communications channel, say with hypertext transfer protocol secure (HTTPS). The following things would typically happen:

1. Alice creates an HTTPS request and sends it to the remote server.
2. The server responds, identifying itself with a cryptographic certificate.
3. Alice's browser checks that the certificate (a) is valid for the original Web address, and (b) has a chain of trust to some well-known and trusted third party, whose public certificate is stored in the browser.
4. Alice and the server are now ready to communicate using the encrypted HTTPS channel for their traffic.

What really happens is illustrated in Figure 7.4. Alice's request goes to Eve's machine. Eve then forwards it on to the server, which uses its own private certificate C1 to create a reply. Eve's machine intercepts this, strips out the server's signature and uses her own private certificate C2 to create a new reply to Alice. Alice receives this and communications are achieved. All traffic from Alice to Eve is encrypted, and all traffic from Eve to the server is encrypted. But Eve now has the ability to decrypt all traffic in either direction, and read information such as passwords.

There is a weakness to this strategy. Eve can construct a certificate C2 that purports to identify the server 10.2.1.43, but it is not likely that she can get it signed by a trusted third party. For example, a company like VeriSign[W] or GeoTrust[X] might have some questions as to why you want a certificate identifying you as, say, PayPal. When Alice connects, her browser will try to warn her, as shown in Figure 7.5.

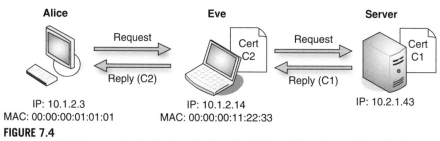

FIGURE 7.4

Eavesdropping Achieved

FIGURE 7.5

You Were Warned

[W]See https://www.verisign.com/index.html.
[X]See http://www.geotrust.com/.

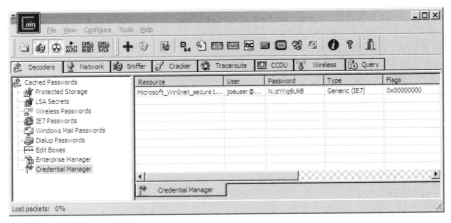

FIGURE 7.6

Password Capture with Cain & Abel

The message "intercept any data you send" is a strong hint as to what is happening. Of course, there are many sites out there with "self-signed" certificates, meaning they use certificates that are not signed by a trusted third party. People may be quick to assume that all is well, and just click the **continue to this website** link. After all, what's the alternative? *Not* checking your bank balance? You need to know! And after all, isn't your bank's security *their* job?

If implementing ARP poisoning sounds very hard to you, take heart. Conversely, if you think it is too hard to worry about it—well, many password sniffers, including Cain & Abel, implement ARP poison routing in a convenient manner.[Y] For Cain & Abel, this comes down to making sure that the tool is set up correctly and then clicking a single button to enable ARP poisoning. In Figure 7.6, Cain & Abel has been used to capture the password "N.izYi!q6UkB" – a very strong password – from an Internet Explorer 7 session.

DANGERS OF PASSWORD REPLAY

Password capture and replay poses a very serious threat to network security, and can be very difficult to guard against because it requires that people choose good passwords, keep track of them in a secure manner, do not fall prey to social engineering schemes, and are vigilant when using secure communications.

One immediate danger is that capturing a password on an otherwise innocuous site such as a personal Web mail provider could lead to compromise of other accounts because of password reuse. For example, corporate policy might require

---

[Y]And dsniff does, too. See the arpspoof tool that is part of the dsniff package. Likewise, Ettercap supports ARP poisoning, as do many other tools.

strong passwords changed relatively frequently. It is unlikely that Rob's Web mail password is the same as his password to the payroll system because he's required to change the latter. However, his VPN might use a certificate secured by the same password he uses for his Web mail. This might allow an attacker to get into his secure corporate e-mail, and from there an attacker might be able to get Rob's payroll password reset. If IT trusts sending passwords in encrypted e-mails that do not pass outside the secure corporate network, this strategy might work.

Replay attacks have dangers of their own because, as with software exploits, you may not know about them as soon as the bad guys do. Worse, if the protocol is itself weak, it may be days, months, or even (as illustrated by the NTLM case described earlier) years before the vulnerability is fixed. Critical infrastructure may depend on the protocol, so just disabling it is not an option.

Many protocols depend on cryptographic hashes for security. As time goes by, these hashes (SHA1, MD5, and so on) are studied and eventually may be cracked. Again, replacing the cryptographic hash at the core of a secure protocol is a nontrivial matter.

DEFENDING AGAINST PASSWORD REPLAY

Several proven technologies exist to avoid password capture and password replay attacks, but one of the most basic ways to resist password theft is to avoid the use of a single, static authentication token whenever possible. Although this is not always possible, it does provide the best means of security.

One method is for the user and the authentication system to augment a shared, static secret like a password with a dynamic, or changing, shared secret. The two must be properly combined in order to gain access. An example of this is the RSA SecurID.[Z] This uses a tamper-resistant device containing a clock synchronized to the server's clock. A shared key is used to generate a sequence of numbers; without the key, the number sequence is nearly impossible to predict from just a few observations. Each number is displayed and is valid for a period of time, and then a new number is generated. When a user wants to log in, they enter a fixed static secret like a password, and combine this with the current number displayed by the SecurID. Someone capturing packets would not be able to replay the authentication later on because the number used would no longer be valid. Even if the attacker got either the user's password or the SecurID device, they would still need the other piece of information to be able to gain entry. Although this is an excellent approach, it can also fall prey to the man-in-the-middle attack. The credentials cannot be permanently compromised, but the secure session can be hijacked using a technique like ARP poisoning.

Another authentication technique is to use one-time passwords. In this case, at login the server generates a challenge. The user looks up the challenge, say on a printed card, and enters the correct response. Each challenge/response pair is good

---

[Z]See http://www.rsa.com/.

for exactly one login. This is a very strong system, and can be combined with a static password so that loss of the card containing the challenge/response pairs would not compromise the system. Again, this can fall prey to a man-in-the-middle attack that hijacks an existing session. A one-time password system that is common on UNIX and Linux systems and is completely software-based is the S/KEY system.[AA]

One reason ARP poisoning attacks work is because layer two of the network does not have any built-in security. Fortunately, there are both software and hardware solutions to this problem. ArpON[BB] is open-source software for detecting and *blocking* ARP poisoning and spoofing attacks, and it runs on Linux and UNIX, including Mac OS X. Antidote[CC] is another open-source ARP poisoning detection system. Several hardware vendors, including Cisco,[DD] have implemented a technique called Dynamic Host Configuration Protocol (DHCP)[EE] "snooping" to detect ARP poisoning or spoofing. Finally, Arpwatch[FF] attacks the problem by watching for ARP messages that reassign an IP address, and generating notifications if this happens.

ArpON can operate in two different ways to defeat ARP poisoning: *static* and *dynamic*. Static ARP inspection works by assuming the ARP cache at program start is valid, and then defending it against modification. This works well if your network consists of machines assigned with static IP addresses. Dynamic ARP inspection works by first clearing the ARP cache, and then carefully monitoring any attempts to modify it and applying rules to prevent ARP poisoning. This works well if your network consists of machines assigned with dynamic IP addresses (DHCP).

THE FUTURE OF PASSWORD REPLAY

In 2004 then Microsoft Chairman Bill Gates, in his keynote address to the RSA Security Conference, predicted the end of traditional passwords.

> *"There is no doubt that over time, people are going to rely less and less on passwords. People use the same password on different systems, they write them down and they just don't meet the challenge for anything you really want to secure."*[1]

A few years have passed, and we're all still using traditional passwords. Bill was right; they are fundamentally flawed as a security measure. It also seems they are not going away any time soon.

There are just too many places you need to authenticate on the Internet. Some means is required; whether it is a password, pass phrase, or combination of questions

---

[AA]The S/KEY One-Time Password System is described in RFC 1760. A free implementation for the Mac is available from http://www.orange-carb.org/SkeyCalc/.
[BB]See http://arpon.sourceforge.net/.
[CC]See http://antidote.sourceforge.net/.
[DD]See http://www.cisco.com/.
[EE]DHCP provides a means to dynamically reserve an IP address for a host, based on the host's MAC address.
[FF]See http://ee.lbl.gov/.

and answers. Your Web designers and your users are all familiar with passwords, and you can't issue special hardware to everyone who registers on your Web site.

Without careful design, protocols are susceptible to replay attacks. Even in cases where protocols *are* designed to be resistant to replay attacks, a weakness in the protocol (as with WEP) can render the protocol susceptible. Finally, even if the authentication portion of the protocol is resistant to replay attacks, it may still be the case that a "man in the middle" can hijack a session and use replay *within* the authenticated session.

Replay attacks are, in some ways, analogous to buffer overflow attacks. They can be eliminated by careful design, and one must keep this in mind when designing an authentication system or protocol. The analogy breaks down, however, when we consider systems with deployed weaknesses. Eliminating a buffer overflow exploit requires shipping a patch. The patch can be tested, and installed on a machine-by-machine basis. For replay vulnerabilities, often the protocol must be *redesigned*. This makes it tough to eliminate the vulnerability, as both endpoints of any potential communication session must be upgraded to compatible, resistant protocols. It may not even be possible to upgrade some legacy systems, as software is no longer being developed for them. Many network endpoints are embedded devices, and the manufacturer may delay in releasing an update, or never release an update at all.

The problem of legacy protocols is often "solved" by allowing one endpoint to downgrade to the old protocol. This is obviously a serious vulnerability. You might have updated your servers to support the newest resistant protocols, but still they have to support older versions of protocols (such as NTLM and SSH-1) because of legacy hardware and software.

Because protocols that are susceptible to replay attacks continue to be designed and deployed, and because weaknesses continue to be discovered in the cryptographic systems that are used to protect against replay attacks, it is clear that replay attacks will remain a very deadly network attack for the foreseeable future.

SUMMARY

After reading this chapter, you should have a better appreciation for the security risks of password capture and password replay. Password capture and replay is a significant, ongoing threat to the security of networks. Because traditional passwords and protocols that are susceptible to replay attacks are not going to go away any time soon, this represents a significant security risk. Further, designing protocols to resist replay attacks requires careful engineering and analysis...so we can assume that even new protocols may be susceptible to replay attacks. Once a vulnerability has been discovered in a protocol, it can be a long time before a fix is available.

Fortunately, there are technologies to help secure against these attacks. The use of one-time passwords, hard-to-guess sequence numbers, and tools like SecurID can block the usual methods of password capture and replay. Sadly, these can still fall prey to man-in-the-middle attacks, made ever easier by well-designed and maintained automated tools. The fundamental message is to evaluate how and when users can authenticate, to establish reasonable policies, and to implement network security auditing. ARP poisoning can be detected on the network using readily available tools. Even so we can expect this network attack to remain deadly for a long time to come.

Endnote

1. See Kotadia M. Gates Predicts Death of the Password, CNET News, February 25th, 2004. Available online at http://news.cnet.com/2100-1029_3-5164733.html; (accessed 2/28/2010).